From Galileo to Lorentz...
and Beyond
Principles of a Fundamental Theory of Space and Time

Joseph Lévy

Apeiron
Montreal

Published by C. Roy Keys Inc.
4405, rue St-Dominique
Montreal, Quebec H2W 2B2 Canada
http://redshift.vif.com

© Joseph Lévy 2003

First Published 2003

National Library of Canada Cataloguing in Publication Data

Lévy, Joseph, 1936-
   From Galileo to Lorentz... and beyond : principles
of a fundamental theory of space and time / Joseph Lévy.

Includes bibliographical references and index.
ISBN 0-9732911-1-7

   1. Space and time. I. Title.

QC173.59.S65L48 2003   530.11   C2003-906045-4

Cover design by Pascale Krier.
Cover graphic: Illustration of the section "On the relativity principle"
in Chapter 2.

The main ideas presented in this manuscript have been registered at the French society of authors.

Table of Contents

Preface i
Acknowledgements ii
Foreword iv
    Steps in the development of the new conception v
Introduction 1
Difficulties with Orthodox Relativity Theory 7
    On light-speed invariance 7
    On the relativity of simultaneity 9
    On the relativity principle 12
    Distance, velocity and time 13
Some Useful Concepts 15
Extended Space-Time Transformations 19
    Introduction 19
    Derivation of extended space-time transformations 22
    Consistency of the transformations 26
    Conclusion 29
    Appendix 1 31
    Appendix 2 33
Inertial Transformations Derived from
Galilean Transformations 39
    Prologue 39
    Inertial transformations between the aether frame
        and other inertial frames 40
    Compatibility of the Galilean law of composition
        of velocities with a limit velocity 44
    Inertial transformations in the general case 44
Arguments in Favour of Lorentz-Fitzgerald Contraction 47
    Introduction 47
    Present-day arguments for Lorentz contraction 50
Implications for Fundamental Physics 59
    The relativity principle 59

    Mass-energy conservation 63
    Principle of inertia 63
    Conservation of momentum 64
    Mass-energy equivalence 64
    Variation of mass with speed 65
    Invariance of the one-way speed of light 65
    Twin paradox 65
    Relativity of time 65
    Relativity of simultaneity 66
    Minkowski space-time 66
    Length contraction 66

Mass and Energy in the Fundamental Aether Theory 69
    Demonstration of $E = mC^2$ without
        relativistic arguments 69
    Variation of mass with speed 70
    Variation of mass with speed in relativity and in
        the fundamental aether theory 72
    The question of reciprocity 75
    Possible measurement of the absolute speed of
        an inertial system 76
    Conservation of energy 76

Synchronization Procedures and Light Velocity 77
    Measuring the speed of light with one or two
        clocks by the Einstein-Poincaré procedure 77
    Measuring the speed of light by the slow
        clock transport procedure 78

References 85

Index 93

Preface

This book draws conclusions from lectures given by the author at various conferences, and from several articles published in collections of essays on relativity theory. Some of the ideas presented here arise from these earlier works, though they may not have been explicitly stated before. The book is inspired in part by the works of Simon Prokhovnik, which represent a fundamental step in the understanding of the theories of space and time. Nevertheless, the reader will appreciate that, beyond a certain point, it departs from Prokhovnik's views.

The reasons for this new course will be explained in detail. We leave the result to the reader's judgment.

The manuscript presents a number of unconventional ideas. We have done our best to refute them, but in the end we find ourselves obliged to accept them as factual. Although we have criticized some of their views, we would like to pay homage to all our predecessors, without whom this manuscript could have not seen the light of day, in particular to Lorentz, Einstein and Poincaré.

3 September 2002

Acknowledgements

We would like to thank all the colleagues who, with an open mind, have helped us to develop and formulate the ideas presented here, in particular:

- The late Prof. Simon Prokhovnik, with whom we kept up a regular correspondence for more than two years. Although the present text criticizes some aspects of Prokhovnik's views, his books, in particular *The logic of special relativity*, were a revelation for us. We feel that they are essential for an understanding of the hidden aspects of space-time theories.
- Prof. Franco Selleri, who has supported some of our ideas and has provided us with the opportunity to contribute to different collections of essays on relativity and quantum mechanics. Prof. Franco Selleri is justly recognized for his prominent contributions in the field of quantum physics and relativity.
- Dr. Michael C. Duffy, organizer of the conference "Physical Interpretations of Relativity Theory" (P.I.R.T), held every two years since 1988 in London and sponsored by the British Society for the Philosophy of Science. We are particularly grateful to Dr Duffy, who, with his characteristic kindness, invited us to deliver talks at this important event. All the scientific community is indebted to Dr Duffy, whose efforts have fostered a renewed interest in questions dealing with relativity and its relationship to other fields of physics.
- The late Prof. Victor Bashkov, organizer of the conference "Geometrization of Physics" held every two years in Kazan University, Russia, for carefully reading and commenting on some of our publications and for inviting us to serve as a member of the International Organizing Committee of the meeting.
- Prof. Paul Marmet, for his comments on the author's book *Relativité et Substratum cosmique* in a review published in *Apeiron*, and for an interesting exchange of views, always done with understanding and friendship.
- Dr. Thomas Phipps Jr, for interesting discussions and exchange of correspondence about convergent views.
- Prof. Jean-Pierre Vigier, for kindly offering to read and comment on the present manuscript and for his prompt reply. It is hardly necessary to say more about Prof. Vigier, whose prominent contribution to modern physics is well known. His support has been for us of the utmost importance.

- C. Roy Keys for his judicious comments and advice. His kind cooperation proved most valuable.

We also wish to thank many of the friends we met at different conferences, in particular at the P.I.R.T meetings (from 1994 to 2000), for inspiring discussions and lectures.

We are indebted to all the colleagues who have offered thoughtful comments on our ideas or provided us with useful information. Among them are Prof. Cavalleri, the late Prof. Grossetete, M.B. Guy, Prof. H. Hayden, Dr R. Hock, Dr P. Huber, M.V. Makarov, M. Marinsek, Dr G. Margalhaes, Prof. M. Sachs, Dr Y. Pierseaux, Prof. J.P. Wesley, and all those who have contributed to our scientific training as a student, such as Prof. A. Abragam, Prof. Brossel, Prof. Cl. Cohen-Tannoudji, Prof. R. Daudel, Prof. R. Lefevre.

Foreword

This text, drawing on recent experimental and theoretical results, develops a fundamental aether theory. By fundamental we refer to the existence of a privileged inertial frame attached to the aether. This implies that two inertial frames, A and B, do not have the same status. There is a hierarchy between the frames. This hierarchical relation conflicts with the idea of relativity.

The text is based essentially on ideas put forward in previous works.[1-4] Nevertheless it does depart from these earlier studies since, in the meantime, it has become apparent to us that certain opinions generally accepted as dogma are, for the most part, questionable.[5,6] But, rather than reject these earlier ideas, we see them as waypoints in the journey that has led us to our present views. The earlier works do shed light on some aspects of these views, and may be helpful to readers who wish to refer to them. However, this is not absolutely necessary since the text is designed to be self-sufficient. (Note that the previous works also deal with subjects that will not be treated here.)

As we will see, Einstein's relativity comes up against a number of objections. As a result, we decided to revisit the approach adopted by Lorentz[8], who assumed length contraction real and non-reciprocal, (in agreement with the views of some classical authors.[9-14])

Nevertheless, *a priori*, some serious objections could also be made to Lorentz's assumptions; for this reason, in previous papers, we did not take a clear-cut position for or against. But, since that time, we have become aware that all the apparent difficulties with Lorentz's assumptions can be overcome. Starting from these assumptions, we were led to the following considerations:

1. When we measure distances, times and speeds in a given inertial frame, we commit systematic errors due to length contraction, clock retardation and unreliable clock synchronization. As a consequence, the experimental space-time transformations that relate two inertial frames connect essentially fictitious data.
2. When one of the two frames is the fundamental frame, by submitting the Galilean coordinates to these distortions we obtain the Lorentz-Poincaré transformations.
3. When we deal with any pair of inertial frames, using the same procedure, we obtain a set of more general transformations that as-

sume a mathematical form different from the Lorentz-Poincaré transformations. These are demonstrated to be consistent; they reduce to the Lorentz-Poincaré transformations in the case cited above, and they allow us to explain why the speed of light, although equal to C exclusively in the aether frame, is always (erroneously) found to be C. They also shed light on some obscure points of physics. The contradiction between the Galilean relationships and the experimental (apparent) space-time transformations is thus removed.

Now, since the form of the space-time transformations is different in the two cases discussed above, they obviously do not obey the relativity principle and do not possess group structure. *A priori* this appeared suspicious to us. But the coherence of this approach, its agreement with recent experimental results, and the fact that it explains why the speed of light is always found to be constant—even though anisotropic—all represented weighty arguments in favour of the theory. The fact that it explains some obscure points of physics which had never been correctly explained before, and some additional arguments, convinced us of its reliability.

In addition, as demonstrated previously,[5,6] and as we will see in this text, there are now strong arguments against the relativity principle.

Of course, this calls into question all the derivations of the Lorentz-Poincaré transformations based on or in agreement with the relativity principle, including one presented in a previous publication.[7] One of the consequences of these results is that the laws of nature should undergo a gradual slow variation as a function of velocity with respect to the aether.

The ideas put forward in this text challenge the theory of special relativity; but they also concern general relativity, which is an extension of the special theory.

Steps in the development of the new conception[*]

It was only by a gradual process that we arrived at the conclusions developed here. Our earlier viewpoints are briefly summarized below:

1) Approaches that call into question the absolute constancy of the speed of light but do not contest other dogmas, such as the relativity principle. These views were expressed in three publications: a brochure entitled "Invariance of light speed reality or fiction," an article by the same title in *Physics Essays*[7] which summarizes the ideas presented in the brochure, and another article entitled "Is the invariance of the speed of light compatible with quantum mechanics?" published in a collection of essays.[24] In these publications, we gave arguments to demonstrate that the

---

[*] Except for some nuances, the terms Cosmic Substratum, aether frame, preferred, fundamental or absolute inertial frame can be considered equivalent.

concept of the absolute constancy of the speed of light cannot be maintained, because it leads to inconsistencies and absurdities. Although our views on the relativity principle have changed, the arguments themselves remain valid.

2) In more recent publications,[1,2,3,4] we became aware that the speed of light was erroneously found to be constant because of three kinds of systematic errors committed in measurements: errors due to length contraction, clock retardation and unreliable clock synchronization.

We were able to demonstrate that the Lorentz-Poincaré transformations can be obtained by exposing the Galilean transformations to these three kinds of errors. Nevertheless, the Lorentz-Poincaré transformations apply exactly only when one of the frames under consideration is the fundamental reference frame in which no aether drift exists. In all other pairs of inertial frames, it was necessary to seek a set of more general transformations, in consequence of which the relativity principle was called into question. Note that in these publications we had not yet found the objections to the relativity principle which were developed in subsequent articles. In addition, length contraction had never been experimentally observed. For these different reasons we could not arrive at a final decision in favour of the Lorentz assumptions.

3) Since that time, we have found a number of arguments which demonstrate that the relativity principle is not indisputable, and that relativity of simultaneity is not essential. The latter concept results from an unreliable synchronization of clocks, which presupposes isotropy of the speed of light in the Earth frame (Einstein-Poincaré procedure[5,6]). These arguments led us to the present approach, which constitutes an alternative theory of space and time where the Lorentz-Poincaré transformations are a particular case of a more general set of transformations. These transformations conceal hidden variables which are nothing more than the Galilean relationships. Although it has never been possible to verify it directly, various arguments lend support to a real, non-reciprocal length contraction. In particular this concept explains why the speed of light is always found to be constant, even though it is actually anisotropic.

Chapter 1

Introduction

> You imagine that, deep down inside, I regard the work of my life with calm satisfaction. The things are really different. There is not a single concept of which I am convinced that it will stand firm.
>
> A. Einstein, Letter to Maurice Solovine, 1949.

The theory of relativity was developed at the end of the nineteenth century and the beginning of the twentieth in order to answer two questions:

1. How to explain why the Galilean transformations did not preserve the invariance of Maxwell's equations under a change of inertial frame.
2. How to justify the fact that measurements of the speed of light always gave the same results, independently of the direction of the signal and of the speed of the source.

At that time, belief in the aether was practically universal. Only a few people speculated about a possible contradiction with the relativity principle.* A number of scientists set out to resolve these problems; they included Heaviside,[9] Fitzgerald,[10] Voigt[17] and Larmor[12]—the principal actors being Lorentz,[11, 16] Poincaré,[18] Einstein[19] and Minkowski,[20] although other lesser-known scientists did play an important role.

Now, it should be borne in mind that the motivations of these authors varied considerably. Lorentz, for example, focussed on the task of finding an explanation for Michelson's experiment that was compatible with the aether hypothesis. He succeeded in this objective by postulating the existence of a length contraction that affected matter in motion with respect to the aether frame. He did not deny the existence of a preferred inertial frame attached to the aether or reject the Galilean law of addition of velocities. In fact, in his explanation of Michelson's experiment, the speed of light is added to or subtracted from the speed of the Earth with respect to the aether frame.

In his early work Lorentz did not show a great interest in the relativity principle. It will be recalled that he did not discover the definitive transfor-

---

* The relativity principle states that the laws of physics must assume the same mathematical form in all inertial frames. (See Chapter 7.)

From Galileo to Lorentz... and Beyond
Joseph Lévy (Montreal: Apeiron 2003)

mations of space and time that bear his name. The transformations he proposed originally were:[16]

$$x' = \beta \ell x \quad y' = \ell y \quad z' = \ell z$$

$$t' = \frac{\ell t}{\beta} - \beta \ell v \frac{x}{C^2} \text{ with } \beta = \left(1 - v^2/C^2\right)^{-1/2}$$

where ℓ is a coefficient which, for small values of v, differs from unity by a quantity of the second order.

It was Poincaré who gave them their classical form and coined the term "Lorentz transformations."[18] (Hence the expression "Lorentz-Poincaré transformations" appears more appropriate and will be used in this text henceforth.)

It is important to note that, although the starting point of Lorentz's work was the Michelson experiment, in which he viewed the speeds as simply additional, the final result was a set of transformations implying a law of composition of velocities different from Galileo's. The reason for this will be discussed later.

Poincaré's aims differed from Lorentz's in several respects. Poincaré was strongly attached to the idea that the laws of nature should assume the same form in all inertial frames. He therefore modified the initial Lorentz transformations so that they would fulfill this requirement.

He extended the Galilean relativity principle to electromagnetism and demonstrated that, insofar as the resulting transformations apply to any pair of inertial frames (a fact we will examine in this text), they assume a group structure.

Nevertheless, Poincaré did not reject the Lorentz aether. This is because, electromagnetic waves needed a medium to propagate, and Poincaré was not familiar with the photon concept, which was Einstein's idea. He expressed his belief in the aether in the following terms:

> Does an aether really exist? The reason why we believe in an aether is simple: if light comes from a distant star and takes many years to reach us, it is (during its travel) no longer on the star, but not yet near the Earth. Nevertheless it must be somewhere and supported by a material medium. (*La science et l'hypothèse*, Chapter 10, p. 180 of the French edition,[21] "Les théories de la physique moderne.")

During a lecture given in Lille (France) in 1909, where he spoke of the new conceptions of modern physics, Poincaré declared:

> Let us remark that an isolated electron moving through the aether generates an electric current, that is to say an electromagnetic field. This field corresponds to a certain quantity of energy localized in the aether (rather than in the electron).

Poincaré acknowledged his debt to Lorentz in the following terms:

> The results I have obtained agree with those of Mr Lorentz in all important points. I was led to modify and complete them in a few points of detail.[18]

Thus, Poincaré tried to reconcile two apparently contradictory notions: the Lorentz aether and the principle of relativity.

Now, as we will see in the following chapters, if we admit the Lorentz aether, we are obliged to realize that the transformations between two inertial frames which bear his name (the Lorentz-Poincaré transformations) apply exactly only when one of the frames (x, y, z, t) or (x', y', z', t') is at rest in the Cosmic Substratum. In all other cases, the space-time transformations assume a different form[1] and, as a consequence, the set of the space-time transformations does not constitute a group.[2] In other words, they do not obey the relativity principle.[2] (See complete demonstration in Appendix 2 of Chapter 4.)

In order for the space-time transformations to obey the relativity principle, it thus seemed necessary to reject the concept of aether and the preferred inertial frame associated with it. This is what Einstein did.[19] Einstein had an advantage over Poincaré: he had discovered the photoelectric effect, which demonstrates the corpuscular aspect of light. Therefore light could, in principle, propagate without the support of a medium, just like any elementary particle. (We now know that Einstein later changed his mind regarding the aether.)

Einstein based his theory on two postulates:[19]

1. Absolute equivalence of all inertial frames for the description of the laws of nature.
2. Invariance of the speed of light.

He then obtained a set of transformations which assume a group structure. The two postulates also implied the concept of relativity of time.

At the time when Einstein developed his theory, all the experimental tests seemed to confirm his views. Measurements of the speed of light performed with increasingly greater accuracy all gave the same value (C). In addition, the covariance of the Maxwell equations predicted by the theory was in agreement with the belief shared by most physicists. In the end, the theory was accepted by virtually the entire scientific community as a dogma.

It was only much later that other physicists realized that the experiments cited in support of Einstein's relativity could be interpreted in a different way.[22] Builder and Prokhovnik, for example, demonstrated that, assuming length contraction, the anisotropy of the one-way speed of light proves compatible with the isotropy of the two-way transit time.*[22] In their argument, they imagined the case of a long rod AB moving uniformly with

---

*This fact has been confirmed repeatedly by modern types of Michelson experiment. See the reference to M. Allais in the "Further References" section at the end of the book.

respect to the aether frame S_0 and making an angle α relative to the direction of motion. The rod was equipped with mirrors at each end in order to allow the reflection of light. The two-way transit time of light along the rod AB was shown to be independent of the angle α and equal to

$$2T = \frac{2\ell_0}{C\sqrt{1-v^2/C^2}} \qquad (1.1)$$

Where ℓ_0 is the length assumed by the rod when it is at rest in the aether frame, and v the speed of the rod with respect to the aether frame. (For a complete demonstration of formula (1.1), consult Chapter 6 "Arguments in Favour of Lorentz-Fitzgerald Contraction" in this text, or the author's monograph *Relativité et Substratum Cosmique.*[3])

Now, as a result of the slowing down of the clocks in the moving frame S (clock retardation) the 'apparent' two-way transit time of light in S is in fact found equal to

$$2\ell_0/C.$$

In addition, the contraction of the rod is not perceived by observer S, since his meter stick is also contracted in the same ratio. As a result, the length of the rod is found equal to ℓ_0 and, in consequence, the average two-way speed of light appears (erroneously) independent of the angle α and of the speed v^*. In other words, it appears isotropic and constant. This is also the case, as will be seen later, for the '*apparent*' one-way speed of light. These questions will be studied in detail in the following chapters.

Important Remarks
1. Note that $2T$ is the real two way transit time of the signal. It is different from the clock reading $2T\sqrt{1-v^2/C^2}$ obtained with clocks attached to the moving frame S. (See Chapter 6.)
2. Contrary to what is often believed, the method of synchronization based on the slow clock transport gives the same result as the Einstein-Poincaré procedure.[3,22] (See Chapter 9.)
3. The De Sitter experiment[23] on the double star beta Aurigae is generally interpreted as proof of the relativity theory; but it can be interpreted in a different way.
   The orthodox explanation given by De Sitter rests on the fact that "light emitted by a source moving with velocity U through space in the direction of motion of the source, has a velocity independent of the motion of the source." De Sitter concluded that the velocity of light is constant.

---

[*] The real average two-way speed of light cannot be isotropic since the two way transit time of light is the same in the two arms of Michelson's interferometer, whereas the two arms have different lengths.

The alternative interpretation replies that, if the speed of light is constant with respect to the aether, it cannot depend, for the receiver, on the speed of the emitter. But this does not mean that it will be constant with respect to the source.

Here we propose:

1. to review some difficulties that afflict orthodox relativity theory
2. to provide a foundation for further developments
3. to revisit the concepts of relativity of simultaneity, relativity principle, and absolute constancy of light velocity
4. to demonstrate that the experimental space-time transformations connect data distorted by systematic errors of measurement. (These errors result from length contraction, clock retardation and imperfect clock synchronization.)

   When these errors are corrected, the hidden variables the transformations conceal turn out to be the Galilean transformations:

$$x = x' + v_0 t$$
$$t = t'$$
$$v = v_0 + v'$$

The Lorentz-Poincaré transformations are a particular case of a set of more general space-time transformations. These are demonstrated to be consistent. Moreover they also shed light on some obscure points of physics.

Note that, as is usually done, the frames associated with bodies not subject to visible external forces are called inertial in this text. But if we assume the existence of an aether drift which exerts a hidden influence on these bodies, they cannot be perfectly inertial. (Only the fundamental frame can be considered truly inertial.)

But, insofar as the aether drift is weak, these frames are almost inertial. This should be the case when their absolute speed does not exceed 10^5 km/sec, since in such cases the ratio m/m_0 falls in the limited range $1 \leq m/m_0 < 1.05$. But at higher speeds, the action of the aether drift should not be ignored.

Chapter 2

Difficulties with Orthodox Relativity Theory

On light-speed invariance

Absolute invariance of light speed comes up against a number of objections. Some of them have already been studied in detail in our book *Relativité et Substratum Cosmique*[3] and in other publications.[7,24] Here we will review these objections and propose a few new ones.

1. If we admit that light is composed of particles, when a photon hits a mirror, its speed must decrease and become zero before it reverses its direction.
2. Absolute invariance of the speed of light also implies a lack of reciprocity between the speed of the source and that of the emitted photons. Now, the principle of reciprocity of relative speeds is a fundamental concept of physics which can be expressed as follows: when a body A recedes from a body B with the speed v, conversely the relative speed of body B with respect to A must be $-v$. Therefore, if we admit that the speed of a photon emitted by a lamp is C, we must also admit that the speed of the lamp with respect to the photon will be $-C$. But this reciprocity is forbidden by the orthodox theory of relativity, because if the lamp had the speed $-C$, its mass would be infinite. The attitude generally adopted in order to resolve this difficulty consists in denying the existence of a proper reference frame for the photon. But this is contrary to logic, because if photons had no frame of reference, it would be impossible to attribute a velocity to them—something we do not hesitate to do.
3. Special relativity implies a proper mass and a proper energy equal to zero for the photon. Nevertheless, since the photon is considered a particle, it is worth asking why it does not possess proper energy (and therefore proper mass). The physicists who conceived special relativity did not answer this question.

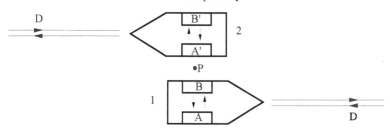

Figure 1. The two rockets paradox.

4. The experiments of Sadeh and Alväger do not prove relativity theory. Let us review these experiments. Sadeh showed that, in the annihilation of positrons with emission of two gamma ray photons, the speed of the photons travelling with a component of motion in the direction of the positron flight was independent of the speed of the source (to within about 10%).[25] Alväger demonstrated that, in the disintegration of π_0-mesons moving at speeds equal to 99.98% C, the velocity of the emitted photons was not affected by the motion of the mesons.[26] These results are equally consistent with aether theories and special relativity. In fact, if the speed of the photons is constant with respect to the aether, it cannot depend, for the receiver, on the motion of the source.

5. Absolute invariance of the speed of light is not compatible with Heisenberg's uncertainty relations.[24]

6. Experiment with two rockets.[3] Consider two rockets receding from one another at constant speed and symmetrically with respect to a point P. Each rocket is equipped with two mirrors A and B (and A' and B') facing one another, at the top and bottom of its fuselage. (See Figure 1.) At instant t_0, the rockets leave point P at speed v, and after covering the distance D they come back to their initial position. Also at instant t_0, a light signal starts from A (and A') and after reflection on B (and B') returns to A (and A'.) The cycle is repeated a number of times until the end of the trip. The traveller in rocket 1, observing the light signal of rocket 2, finds that the path of the ray is longer than his own. Assuming that the speed of light is constant, he concludes that a cycle in rocket 2 takes longer than in rocket 1. Since, according to relativity, nothing differentiates between the trips, the number of cycles completed by the light ray will be the same in the two rockets; but, as the cycles in rocket 2 appear longer, observer 1 will paradoxically conclude from his calculation that the duration of the trip by rocket 2 is longer. Conversely, observer 2 will draw opposite conclusions, as can be seen in Figure 2. Obviously the points of view of the two travellers are contradictory and cannot be true at the same time. In fact, the only valid conclusion is that the duration of the two trips is identical. And, as a consequence, we must realize that

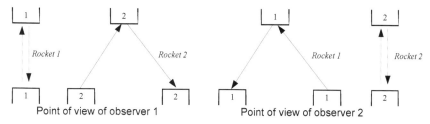

Figure 2. Contradictory points of view of observers 1 and 2.

the speed of light is not the same for an observer inside a rocket and another outside the rocket. Therefore, if the speed of light is found equal to C independently of α and v, this is because of the erroneous measurements mentioned above.

On the relativity of simultaneity

Let us briefly recall the arguments of special relativity and offer a succinct reply. The question has also been studied in previous publications.[4,6] According to special relativity, two distant events which are simultaneous for one observer, are not simultaneous for another observer moving with respect to the first with uniform rectilinear motion.

In order to demonstrate this theorem, Einstein takes the classical example of the train and the two flashes of lightning[27] (Figure 3): two flashes strike at the two ends A and B of an embankment at the very instant when the middle O' of the train meets the middle O of the embankment (Einstein).

By definition, the two flashes will be considered simultaneous with respect to the embankment if the light issuing from the flashes meets the middle of the embankment at the same instant. Einstein adds that the definition is also valid for the train, but as the train is travelling toward point B, the light coming from B will reach the middle of the train before the light coming from A. Einstein concludes that two events simultaneous for the observer standing on the embankment are not simultaneous for the observer in the train.

As we have suggested elsewhere,[4] Einstein's definition is only appropriate if the light sources are firmly tied to the reference frame in which the measurement is carried out. In the example here, if the light is emitted by

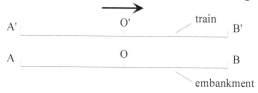

Figure 3. At the initial instant, the middle of the train coincides with the middle of the embankment.

two lamps attached to the embankment, the definition will be true for the embankment, but not for the train, and *vice versa*. Einstein himself actually recognizes that the light coming from B must cover a shorter path to reach O' than the light coming from A. But the definition is only valid if the light covers the same path in both directions. (Note that the Earth also moves with respect to the train.)

If we are to reason correctly, we must bear in mind that simultaneous reception of signals does not necessarily imply simultaneous emission.

Hence, to accurately define the simultaneity of two events we must specify the following: Two instantaneous events, occurring at two points A and B and emitting light in opposite directions toward a point O which is the middle of AB at the beginning of the experiment, can be considered simultaneous if the light emitted from A and B reaches point O at the same instant, provided that O remains fixed with respect to A and B all through the experiment.

Of course, this definition is exact only if the speed of light is equal in both directions. In the fundamental aether theory the definition is considered exact in the aether frame only, and not in other frames, since in those frames the one-way speed of light is not isotropic.

We will now propose another definition of simultaneity that is valid when point O' moves with respect to A and B. Let us reconsider the example of the train and the embankment just seen. (See Figure 4.) Contrary to the author's earlier study[4] we will not suppose *a priori* that the speed of light is identical in both directions: so we will designate the speed of light in the two opposite directions C_{AB} and C_{BA}.

We now place three perfectly synchronized clocks at A, O and B. At the initial instant, O coincides with O'. At this very instant two signals are sent from A and B in opposite directions. When the signal emitted from A reaches point O, point O' has moved towards B a distance $\ell_0 v/C_{AB}$, where v is the speed of the train and $\ell_0 = AO$. When the signal has covered this distance in turn, point O' has moved a distance

$$\frac{v}{C_{AB}}\left(\frac{v}{C_{AB}}\ell_0\right),$$

and so on. Thus, in order to reach point O' the signal must cover the distance

$$\ell_0\left(1+\frac{v}{C_{AB}}+\frac{v^2}{C_{AB}^2}+\ldots+\frac{v^n}{C_{AB}^n}+\ldots\right)=\ell_0\frac{C_{AB}}{C_{AB}-v},$$

and the time needed to cover this distance will be

$$t_{AO'}=\frac{\ell_0}{C_{AB}-v}.$$

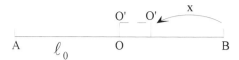

Figure 4. When the light signal coming from B reaches the middle of the train, the train has covered the distance OO'.

Now, in order to reach the middle of the train, the signal emitted from B will cover a distance x, such that

$$\text{OO'} = \frac{v}{C_{BA}} x = \ell_0 - x. \text{ (See Figure 4.)}$$

Therefore

$$x = \frac{\ell_0 C_{BA}}{C_{BA} + v};$$

and the time needed to cover the distance will be:

$$t_{BO'} = \frac{\ell_0}{C_{BA} + v}.$$

Hence, we can conclude that two instantaneous events occurring at A and B can be considered simultaneous if the light emitted by them reaches the middle of the train at two instants $t_{AO'}$ and $t_{BO'}$, such that

$$t_{AO'} - t_{BO'} = \ell_0 \left(\frac{1}{C_{AB} - v} - \frac{1}{C_{BA} + v} \right).$$

Important remarks

1. In our previous article[4] we assumed that the speed of light was isotropic, such that $C_{AB} = C_{BA}$. This is true exclusively in the fundamental inertial frame. In this case we have

$$t_{AO'} - t_{BO'} = \frac{2\ell_0 v}{C^2 \left(1 - v^2/C^2\right)},$$

where C is the speed of light in the preferred frame.

2. Note that in this example, $t_{AO'}$ and $t_{BO'}$ are the real times given by the clocks in the privileged frame. As a result of clock retardation, the clocks in the train will display different times. But this will not affect our reasoning or our conclusions regarding the absolute character of simultaneity. The same remark can be made with regard to the length contraction affecting the train.

Another example that confirms the absolute character of simultaneity

Figure 5 An example of the absoluteness of simultaneity.

will certainly help to convince the wavering reader. Consider two rigid collinear rods AB and A'B' moving toward one another uniformly along the same line. The rods are assumed to be of identical length when they are in motion with relative speed v. They are firmly fixed to reference frames S and S' respectively. (See Figure 5.) Two identical perfectly synchronized clocks are placed in A and B. Another pair of such clocks are placed in A' and B'. In order to reach A', clock A must cover a distance $D = AA'$. Clock B must cover the same distance in the direction of B' (since $BB' = AA' = D$).

Therefore, for an observer in frame S, the encounters between A and A' and B and B' will be simultaneous. But an observer in frame S' will draw the same conclusions. Hence the two observers conclude that both events are simultaneous. (Note that this does not imply that the clocks in frame S will show the same time as those in S'. The conclusion concerns simultaneity and not the clock display. We must distinguish between clock retardation and relativity of simultaneity.)

We must be aware that an apparent relativity of simultaneity exists. It is inherent in the Lorentz-Poincaré transformations and the extended space-time transformations that will be studied later. But it is not essential, and results from the synchronism errors introduced by the Einstein-Poincaré synchronization procedure (or by the method of slow clock transport). These methods of synchronization are not ideal but they are the simplest and most often used. When the systematic errors of measurement entailed by these methods are corrected, the absolute character of simultaneity is recovered.

A criterion of absolute simultaneity was given in a previous paper.[6] let two identical rubber balls bounce on the two pans of a precision balance. If the central pointer of the beam does not move, we can assert that the balls bounced at the same instant. This is true for all observers, at rest or in motion with respect to the balance, whether they are accelerated or not. Of course, a small correction would be necessary, since there could be a minute difference in the speed of propagation of the vibration along the two arms of the beam of the balance. But this is of no consequence, since the correction would be identical for all observers.

Note that the notion of four dimensional space is also conventional, and once errors of synchronization are corrected, the mixing of space and time disappears.[5,6] This claim will be demonstrated later.

On the relativity principle

According to Poincaré's relativity principle, it is impossible by means of an experiment internal to a given inertial frame, S, to determine whether the frame is at rest or in motion with respect to the aether frame. This statement is questionable, as the following demonstration will show: consider two vehicles moving uniformly in opposite directions along a straight line. At the initial instant t_0, they meet at a point O of frame S and then continue on their way symmetrically at speed v toward two points A and B situated at equal distances from point O. At the instant they meet, the clocks inside the vehicles are synchronized. According to Poincaré's relativity principle, the clocks should display the same time when they reach points A and B. If they did not, this would represent a way of determining if frame S is at rest or in motion with respect to the aether frame.

But, insofar as the vehicles do not move at the same speed with respect to the aether frame, the slowing down of their clocks with respect to the clocks at rest in this preferred frame will be different, and so they will display different times. (The clock displays would be identical only if frame S were at rest in the Cosmic Substratum.)

Thus, the relativity principle appears incompatible with the existence of an aether frame in a state of absolute rest[4,5] (contrary to Poincaré's approach). Since a number of arguments now support the existence of this aether frame, the relativity principle must be discarded.

This experiment would, in principle, make it possible to synchronize the clocks placed at A and B. To accomplish this, the clocks need only be synchronized with the clocks inside the vehicles when the vehicles pass in front of them. If the clocks in the vehicles display different times when they pass in front of A and B, one of the clocks A or B must then be adjusted so as to display the same time as the other.

Distance, velocity and time

In his original formulation of special relativity, Einstein denied the concept of aether. Later he changed his mind in order to formulate the general theory of relativity. In his booklet *Sidelights on Relativity* he expressed his views in the following terms:

> ...according to the general theory of relativity, space is endowed with physical qualities; in this sense therefore there exists an aether... But this aether may not be thought of as endowed with the quality of ponderable media, as consisting of parts which may be tracked through time. The idea of motion may not be applied to it.

It is therefore clear that for Einstein one cannot assume an aether frame with a definite position in space, since this aether would be in motion with respect to any other inertial frame.

As we will see in that which follows, Einstein's conception is unacceptable. The argument for its rejection was developed in a previous paper.[5] Here we reformulate it in a different way and make further remarks.

The argument rests on the following three propositions:

1. The real relative speed v between two bodies A and B receding uniformly from one another along the same line is invariant. It is the same for A and for B.
2. The real relative distance does not depend on which one measures it.
3. Consequently, the time t needed for the bodies to recede from each other from distance zero to distance ℓ is also invariant.

For example, consider a meson that travels the distance ℓ from the Earth to a point P and then decays. Its life-time must be the same for an observer on Earth as for an observer moving with the meson.

At very low speeds ($v \cong 0$) with respect to the Earth, the life-time of the meson is, for example, t. At high speed measured from the Earth frame, suppose that we find T.

According to relativity, t is the proper life-time of the meson, and since, in this theory, there is no difference between rest and uniform motion, the proper life-time of the meson at high speed (measured by an observer at rest with respect to the meson) must also be t. Nevertheless, if this were true, it would contradict propositions 1, 2 and 3.

We can therefore conclude that, contrary to relativity, the proper life-time of the meson at high speed is also T (and it is different from the life-time at low speed, which is t.)

Actually if the proper life-times at rest and at high uniform speed are different, this is because there is a difference between rest and motion, which implies that motion possesses an absolute character. This conclusion is at variance with the relativity principle. In other words, the rest frame and the moving frame are not equivalent.

Apparent coordinates

The life-time of the meson at high speed is T. But clocks moving with the meson (at rest with respect to it) would display approximately the time

$$t \cong T\sqrt{1 - v^2/C^2}$$

Therefore, t is not the real time, but the clock display due to clock retardation. This result is only approximate, because the velocity of the Earth with respect to the Cosmic Substratum is not zero. But this velocity is very small compared to the speed of high energy mesons, which approximates the speed of light.

Chapter 3

Some Useful Concepts

Proof that the Lorentz-Poincaré transformations connect the real co-ordinates of the aether frame with the apparent (fictitious) co-ordinates of any other inertial frame. (Example of a light signal.)

Consider two reference frames, one of them (S_0) at rest in the Cosmic Substratum, and the other (S') moving at speed v along the x-axis of frame S_0, with rectilinear uniform motion.

At the initial instant, the two frames are coincident. At this very instant, a light signal starts from the common origin O, O' and travels along a rod O'A firmly tied to frame S' and aligned along the x'-axis. (See Figure 6.) According to Lorentz, the rod O'A is contracted in such a way that

$$\ell = \ell_0 \sqrt{1 - v^2/C^2},$$

where ℓ_0 is the length of the rod when it is at rest in frame S_0, ℓ is the contracted length, and C the speed of light in frame S_0.

Let us determine the time needed by the signal to reach point A. We remark that, when the signal has covered a distance in frame S_0 equal to ℓ, frame S' has moved a distance $\ell v/C$ away from S_0. When the light has covered this latter distance, frame S' has moved away a further distance equal to

$$\frac{v}{C}\left(\frac{v}{C}\ell\right),$$

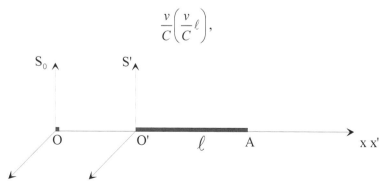

Figure 6 - S' moves uniformly with respect to S_0 along the x-axis.

From Galileo to Lorentz… and Beyond
Joseph Lévy (Montreal: Apeiron 2003)

and so on, so that, in order to reach point A, the signal must cover the distance

$$\ell\left(1 + \frac{v}{C} + \frac{v^2}{C^2} + \ldots + \frac{v^n}{C^n} + \ldots\right).$$

The sum of the series is

$$\begin{aligned}
x &= \ell\frac{C}{C-v} \\
&= \ell_0\sqrt{1-v^2/C^2}\,\frac{C}{C-v} \\
&= \frac{\ell_0 + v\ell_0/C}{\sqrt{1-v^2/C^2}},
\end{aligned} \tag{3.1}$$

and the time needed to cover this distance will be:

$$\begin{aligned}
t &= \frac{\ell_0\sqrt{1-v^2/C^2}}{C-v} \\
&= \frac{\ell_0/C + v\ell_0/C^2}{\sqrt{1-v^2/C^2}}.
\end{aligned} \tag{3.2}$$

Now, assuming that the speed of light in frame S' is $C - v$ and that the clocks used to measure the time are perfectly synchronous, an observer in frame S' will obviously find for the time (in the absence of clock retardation)

$$t = \frac{\ell_0\sqrt{1-v^2/C^2}}{C-v},$$

which is identical to formula (3.2). As a result, the time needed to cover the distances separating the origins of the frames and point A appears identical in both frames.

Yet the problem is not so simple as it appears at first sight, because in order to measure the transit time of light in frame S' from O' to A, we must synchronize clocks beforehand. To accomplish this we generally use the Einstein-Poincaré (E.P) method, which assumes that the speed of light is identical in any direction of space (or the slow clock transport method, which is approximately equivalent.[3,22] See Chapter 9.) We must therefore send a light signal from O' to A. After reflection on A, the signal comes back to O'.

Let us designate the *apparent* two-way transit time of light (measured with clocks in frame S') as $2T$, and the instant of departure of the signal as $t_{O'}$. According to the E.P procedure, since the speed of light is assumed isotropic, the signal is supposed to reach point A at time t_A such that

$$t_A = t_{O'} + T.$$

Now if we suppose, as Lorentz did, that the speed of light is not isotropic in frame S', but instead equals $C-v$ in the direction O'A and $C+v$ in the opposite direction, then the average transit time of the signal along the distance O'A will be:

$$\tau = \frac{1}{2}\ell_0\sqrt{1-v^2/C^2}\left(\frac{1}{C-v}+\frac{1}{C+v}\right)$$
$$= \frac{\ell_0}{C\sqrt{1-v^2/C^2}}. \tag{3.3}$$

Of course, for Lorentz, τ is not the one-way transit time of light, but as we have seen, the Einstein-Poincaré procedure does not make allowance for the anisotropy.

Now, if we take clock retardation in S' into account, formula (3.3) has to be multiplied by $\sqrt{1-v^2/C^2}$. The *'apparent'* experimental (fictitious) one-way transit time of light t'_{app} along O'A in frame S', will then be found equal to

$$t'_{app} = \tau\sqrt{1-v^2/C^2} = \frac{\ell_0}{C},$$

and since the meter stick used to measure ℓ_0 is contracted in the same ratio as the rod, the apparent length of the rod x'_{app}, will be found equal to ℓ_0, so that

$$x'_{app} = \ell_0.$$

Inserting these results in (3.1) and (3.2) we obtain the Lorentz-Poincaré transformations:

$$x = \frac{x'_{app}+v\,t'_{app}}{\sqrt{1-v^2/C^2}},$$

$$t = \frac{t'_{app}+vx'_{app}/C^2}{\sqrt{1-v^2/C^2}}.$$

We conclude that the Lorentz-Poincaré transformations connect the real times and distances in frame S_0, with the experimental (but fictitious) times and distances in frame S'. (Even though they are fictitious, these times and distances are useful because they are the experimental results, and they will help us to find the real distances and times in frame S'. See below.)

Synchronism discrepancy effect (SDE)

The synchronism discrepancy effect (or anisotropy effect) as defined by S. Prokhovnik, is equal to the difference between the time displayed by a clock taken as a reference system and that of another clock synchronized with the first by means of a definite method we want to test.

If the synchronization procedure were perfect, the SDE would be zero.

If we use the Einstein-Poincaré procedure, the synchronism discrepancy between the clocks O' and A in the system discussed above (Figure 6) is equal to the difference between the real time needed by a light signal to go from O' to A and the apparent time (*i.e.*, the average transit time).

In the absence of clock retardation (that is, if the clocks in the moving frame worked at the same rate as the clocks in the aether frame), according to formulas (3.2) and (3.3) this effect would be equal to

$$\delta = t - \tau = \frac{v\ell_0}{C^2\sqrt{1-v^2/C^2}}, \qquad (3.4)$$

but since the clocks in the moving frame are slowed down, the apparent SDE in frame S' is given by

$$\Delta = \delta\sqrt{1-v^2/C^2} = \frac{v\ell_0}{C^2}. \qquad (3.5)$$

As we will see in Chapter 6, this consideration (which had escaped Prokhovnik) is important and leads to a full explanation of why the speed of light (even though it is anisotropic) is always found to be constant.

Chapter 4

Extended Space-Time Transformations

Introduction

The theory of space-time transformations came about as the result of collective work by a number of physicists. Although the fact is often ignored, there were two opposing currents of thought. The relativistic viewpoint has prevailed for a long time but, today, convincing experimental and theoretical arguments lend support to the alternative viewpoint which assumes the existence of a fundamental inertial frame.

The purpose of the present chapter is to re-examine the Lorentz assumptions in light of this new data, and to verify whether they are in agreement with various well established concepts of physics. If they are found to be in agreement, we would be justified in seeking a set of transformations derived from the Lorentz postulates, but more general than those of Lorentz-Poincaré and which could be applied between any pair of inertial frames. (We should bear in mind that the Lorentz postulates are: existence of a preferred inertial frame attached to the aether, called Cosmic Substratum, length contraction, clock retardation, variation of mass with speed, speed of light isotropic and equal to C exclusively in the aether frame.)

At this point, it would be appropriate to briefly review some of the notions taken up in our previous papers that will be developed here. In those publications[1-4] we demonstrated, in the special case of a light signal emitted from an inertial frame distinct from the fundamental frame, that the space-time transformations connecting this reference frame and any other frame take a form different from the classical Lorentz-Poincaré transformations. These transformations therefore do not constitute a group and, consequently, do not obey the relativity principle. (See Appendixes 1 and 2 of the present chapter.)

We also demonstrated that the Galilean composition of velocities law can be easily converted into the law $v = (v_0 + v')/(1 + v_0 v'/C^2)$ by introducing the systematic errors of measurement resulting from length contraction, clock retardation and unreliable clock synchronization.[2] (Note that the law takes this form exclusively when one of the frames is the fundamental frame. In all other cases, as we will see later, it takes another form.)

From these considerations, in agreement with experimental facts, we concluded that the space-time transformations derived from the Lorentz postulates, which connect two inertial frames, possess the following properties:

1. They are reducible to the Galilean transformations (See Chapter 1.) after correction for the erroneous measurements mentioned above.
2. They assume their usual form (Lorentz-Poincaré transformations) only when one of the frames is the preferred inertial frame (aether-frame.)
3. They are in contradiction with the relativity principle. (This property is a direct consequence of the second property.)

Initially, these observations caused us to be suspicious of the Lorentz assumptions. Indeed, in order to demonstrate the law $m = m_0 \gamma$, we generally make use (as Einstein did) of the law of conservation of total relativistic momentum in any inertial frame, which applies in relativity theory. If Lorentz's theory is reducible to Galileo's, this conservation law cannot be used to demonstrate $m = m_0 \gamma$. Thus, at first sight, Lorentz's other assumptions seemed incompatible with one of them $m = m_0 \gamma$, which is an important experimental law. Finally, the fact that the Lorentz assumptions are not compatible with the relativity principle seemed to constitute an important objection to this approach.

But, since that time, we have become aware that the relativity principle is not an indisputable concept of physics.[5] (See Chapters 2, 4, 7 and 8.) As a result, it cannot really be used to demonstrate or refute the law $m = m_0 \gamma$. Indeed, if the laws of physics are not perfectly invariant, then there is no need for the total relativistic momentum to be exactly conserved in all inertial frames. Thus, an important objection to Lorentz's assumptions was removed. This result can be at least partly explained if we assume that the absolute speed of particles interacting in a collision is reduced by the aether drift. In that case, the total relativistic momentum of these particles cannot be exactly the same before and after the collision. This effect cannot be ignored at high speeds ($v \geq 10^5$ km/sec). (Note that for Wesley,[30] the conservation law is strictly valid in the fundamental frame only.)

In Chapter 8, using arguments independent of relativity, we will demonstrate that the law $m = m_0 \gamma$ is in fact compatible with the other Lorentz assumptions.

Of course the mass-energy conservation law remains universal, but we must make allowance for exchanges of energy with the aether, particularly at high speeds, $v > 10^5$ km/sec. (See important remark after formula (4.9).)

However, certain points remained to be clarified:

1. If the Lorentz-Poincaré law of composition of velocities is reducible to the Galilean law, is it compatible with the existence of a limit velocity?
2. Are there any experimental facts that render length contraction compulsory?
3. In the event experimental support exists, is length contraction necessary to explain other experimental results, such as the *apparent* (measured) isotropy of the two-way speed of light?

We will endeavour to answer these three questions.

Let us suppose that we have two inertial bodies, one travelling at the speed $v_0 = 2 \times 10^5$ km/sec with respect to the Earth, and the other at $v = 2 \times 10^5$ km/sec with respect to the first, the two bodies and the Earth being aligned.

According to Einstein's special relativity, the speed of the second body with respect to the Earth (v') is

$$v' = \frac{2v}{1+v^2/C^2} = \frac{4 \cdot 10^5}{1+4/9} = \frac{36}{13} 10^5 \text{ km/sec} < C.$$

In Galilean theory, the total speed would be $v' = 2v = 4 \times 10^5$ km/sec $> C$, which is not in agreement with experiment. But this obstacle can be overcome if we simply assume that the speed of a body with respect to the aether frame (V) is limited in such a way that $V < C$. This means that, if a body A moves away at speed v_A from the origin of a co-ordinate system at rest with respect to the aether frame, the speed relative to A of another body B moving along the direction OA will be limited to

$$v_B < C - v_A.$$

The answers to questions 2 and 3 will be given in detail in Chapter 6.

In light of this new data, the Lorentz assumptions appear far better founded today than in the past. Nevertheless, as will be seen later, this statement applies to the Lorentz assumptions, but not the Lorentz-Poincaré transformations. It should be borne in mind that Lorentz-Poincaré's transformations apply only in a particular case.[1-3]

Here we propose to derive a set of space-time transformations that are valid in all inertial frames. This approach is parallel to, but different from Selleri's work,[28] since, here, we are dealing with the experimental transformations obtained from the usual Einstein-Poincaré method of synchronization, or from the slow clock transport method, which is almost equivalent, but not with the absolute synchronization procedure which would be difficult to apply experimentally.

In addition, we will show that the co-ordinates used in these transformations, although experimental, are fictitious and must be corrected in order to obtain real lengths, speeds and times. Conversely, in order to derive them, we must make use of the Galilean transformations. Finally, these extended space-time transformations apply to all moving bodies and not exclusively to the special case of a light signal, which was studied in earlier work.[1]

Derivation of extended space-time transformations

In order to derive the transformations, we will use the following thought experiment:

Consider three inertial systems S_0 (O, x_0, y_0, z_0), S_1 (O', x_1, y_1, z_1) and S_2 (O", x_2, y_2, z_2). S_0 is at rest in the Cosmic Substratum (aether frame,) S_1 and S_2 are moving along the common x-axis with uniform rectilinear motion (See Figure 7.) We propose to derive the space-time transformations between frames S_1 and S_2.

A long rigid rod O"A at rest with respect to frame S_2 is aligned along the x_2 axis. At the initial instant, the origins of the three frames O, O' and O" are coincident. At this instant, a moving body M (coming from the $-x_2$ region) passes by O", and then continues on its way along the rod with uniform rectilinear motion toward point A.

We will designate the speed of reference frame S_1 with respect to S_0 as v_{01}, the speed of S_2 with respect to S_0 as v_{02}, and the speed of S_2 with respect to S_1 as v_{12}. The speed of the body with respect to S_0 will be called V, (with $V > v_{02}$.) The length of rod O"A would be ℓ_0 if frame S_2 were at rest with respect to the Cosmic Substratum. But, as a result of its motion, O"A is contracted, and assumes the length ℓ, such that

$$\ell = \ell_0 \sqrt{1 - v_{02}^2/C^2} .$$

When the body reaches point A, it meets a clock $C\ell$ equipped with a mirror, firmly tied to frame S_1 and standing at a point A' in this frame (so that when the body arrives at point A, A and A' are coincident).

The real distance X_{1r} covered by the body in frame S_1 can easily be obtained: indeed, the ratio of the distances covered in frames S_1 and S_2 is equal to the ratio of the speeds with respect to these two frames,

$$\frac{O'A'}{\ell_0\sqrt{1 - v_{02}^2/C^2}} = \frac{V - v_{01}}{V - v_{02}},$$

leading to

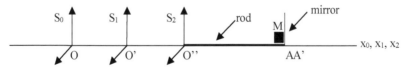

Figure 7. When the body arrives at point A, it meets a clock equipped with a mirror firmly attached to frame S_1 at a point A' in this frame.

$$O'A' = X_{1r} = \ell_0 \sqrt{1 - v_{02}^2/C^2} \frac{V - v_{01}}{V - v_{02}}. \tag{4.1}$$

Notice that, as we have seen in Chapter 3, real speeds are simply additive. On the contrary, as we will now see, only the *apparent* speeds obey a law of composition of velocities different from the Galilean law.

Since the distance O'A' is measured with a contracted meter stick in frame S_1, the *apparent* distance X_{1app} found by observer S_1 will be:

$$X_{1app} = \frac{\ell_0 \sqrt{1 - v_{02}^2/C^2}}{\sqrt{1 - v_{01}^2/C^2}} \frac{V - v_{01}}{V - v_{02}}. \tag{4.2}$$

Now, in order to determine the time needed by the body to reach point A', we must first synchronize two clocks placed at O' and A'.

As we saw in Chapter 3, the Poincaré-Einstein method treats the clock display,

$$\tau_{1app} = \frac{t_1 + \bar{t}_1}{2} \sqrt{1 - v_{01}^2/C^2},$$

as the one-way transit time of light. However, this is actually the *'apparent'* average transit time of light from O' to A' and to O' again. The real transit time of light from O' to A' is in fact:

$$t_1 = \frac{X_{1r}}{C - v_{01}}, \tag{4.3}$$

and from A' to O':

$$\bar{t}_1 = \frac{X_{1r}}{C + v_{01}}. \tag{4.4}$$

When clock retardation in frame S_1 is taken into account, the synchronism discrepancy Δ between the clocks placed at O' and A' is given by

$$\Delta = t_1 \sqrt{1 - v_{01}^2/C^2} - \frac{(t_1 + \bar{t}_1)}{2} \sqrt{1 - v_{01}^2/C^2}$$
$$= \frac{(t_1 - \bar{t}_1)}{2} \sqrt{1 - v_{01}^2/C^2}.$$

From (4.1), (4.3), and (4.4) we obtain:

$$\Delta = \frac{v_{01} \ell_0}{C^2} \frac{\sqrt{1 - v_{02}^2/C^2}}{\sqrt{1 - v_{01}^2/C^2}} \frac{V - v_{01}}{V - v_{02}}.$$

Now the real time needed by the body to cover the distance O'A' in frame S_1 is

$$T_{1r} = \frac{X_{1r}}{V - v_{01}} = \frac{\ell_0 \sqrt{1 - v_{02}^2/C^2}}{V - v_{02}} \quad \text{(from (4.1))}.$$

This time is the universal time measured with clocks attached to frame S_0, in which there is no speed of light anisotropy and no clock retardation. However, in frame S_1 we must take the synchronism discrepancy effect and clock retardation into account, so that the experimental 'apparent' time obtained when we use the Poincaré-Einstein method is:

$$T_{1app} = T_{1r}\sqrt{1-v_{01}^2/C^2} - \Delta$$

$$= \ell_0 \frac{\sqrt{1-v_{02}^2/C^2}}{\sqrt{1-v_{01}^2/C^2}} \frac{(1-v_{01}V/C^2)}{V-v_{02}}. \tag{4.5}$$

From expressions (4.2) and (4.5) we obtain

$$V_{1app} = \frac{X_{1app}}{T_{1app}} = \frac{V-v_{01}}{1-v_{01}V/C^2}. \tag{4.6}$$

Equations (4.2) and (4.5) can be expressed as functions of T_{2app} and X_{2app}. Note that, since ℓ_0 is measured in frame S_2 with a contracted meter stick, we find $X_{2app} = \ell_0$. We also note that

$$X_{2app} = \frac{V-v_{02}}{1-v_{02}V/C^2} T_{2app}.$$

Replacing ℓ_0 in (4.5) with this expression we obtain

$$T_{1app} = T_{2app} \frac{\sqrt{1-v_{02}^2/C^2}}{\sqrt{1-v_{01}^2/C^2}} \frac{(1-v_{01}V/C^2)}{(1-v_{02}V/C^2)}, \tag{4.7}$$

and replacing ℓ_0 with X_{2app} in (4.2) we have

$$X_{1app} = X_{2app} \frac{\sqrt{1-v_{02}^2/C^2}}{\sqrt{1-v_{01}^2/C^2}} \frac{V-v_{01}}{V-v_{02}}. \tag{4.8}$$

Conversely

$$T_{2app} = T_{1app} \frac{\sqrt{1-v_{01}^2/C^2}}{\sqrt{1-v_{02}^2/C^2}} \frac{(1-v_{02}V/C^2)}{(1-v_{01}V/C^2)},$$

$$X_{2app} = X_{1app} \frac{\sqrt{1-v_{01}^2/C^2}}{\sqrt{1-v_{02}^2/C^2}} \frac{V-v_{02}}{V-v_{01}}.$$

We can now see that, contrary to special relativity, v_{01} and v_{02}, which are the velocities of frames S_1 and S_2 with respect to the aether frame, are systematically omnipresent in the equations.

Expressions (4.7) and (4.8) are the extended space-time transformations, applicable to any pair of inertial bodies receding uniformly from one another along the direction of motion of the Solar system. It should be pointed out that, during a very brief interval, the motion of the Earth with respect to the Cosmic Substratum can be considered rectilinear and uniform. If this were not the case, the bodies placed on its surface would be subjected to perceptible accelerations.

Compared to the absolute motion of the sun, the orbital and rotational motions of the Earth are very slow. Thus, as a first approximation and for a short time, the motion of the Earth can be identified with that of the solar system. This means that S_1 can be identified with the Earth frame, and S_2 with a vehicle moving on its surface in the direction of motion of the Solar system (a ship for example). Accordingly, v_{01} is the real speed of the Earth with respect to the Cosmic Substratum (S_0); v_{02} is the real speed of the ship with respect to S_0; V is the real speed, with respect to S_0, of a body present on the surface of the ship and moving in the same direction; and V_{1app} is the apparent speed of the body with respect to the Earth frame.

Formula (4.8) is identical to the expression of the space transformation given in a previous publication.[66] Formula (4.7) represents a completely satisfactory expression of the time transformation: it applies to all values of V. It replaces the expression given in the earlier publication, which was limited to high values of V.

It is interesting to express V_{1app} as a function of V_{2app}. From $V_{2app} = (V - v_{02})/(1 - v_{02}V/C^2)$ we obtain

$$V = \frac{V_{2app} + v_{02}}{1 + v_{02}V_{2app}/C^2}.$$

Substituting this expression for V in (4.6) we obtain

$$V_{1app} = \frac{\dfrac{V_{2app} + v_{02}}{1 + v_{02}V_{2app}/C^2} - v_{01}}{1 - \dfrac{v_{01}}{C^2}\left(\dfrac{V_{2app} + v_{02}}{1 + v_{02}V_{2app}/C^2}\right)}.$$

Conversely, from

$$V = \frac{V_{1app} + v_{01}}{1 + v_{01}V_{1app}/C^2},$$

we easily find

$$V_{2app} = \frac{\dfrac{V_{1app} + v_{01}}{1 + v_{01}V_{1app}/C^2} - v_{02}}{1 - \dfrac{v_{02}}{C^2}\left(\dfrac{V_{1app} + v_{01}}{1 + v_{01}V_{1app}/C^2}\right)}.$$

We remark that, in conformity with the *'apparent'* (experimental) speed of light invariance,

$$V_{1app} = C \Rightarrow V_{2app} = C, \qquad (4.9)$$

and *vice versa*.

Important remark

It is commonly asserted that the total relativistic momentum of particles interacting in a collision is conserved even when the collision occurs at very high speed. This appears highly doubtful. Indeed, when clocks are synchronized by the usual methods, systematic errors are made in measuring the speeds, and apparent speeds (V_{app}) are found in place of real speeds.

Consistency of the transformations

The consistency of the transformations can now be checked. To this end, we need to demonstrate their agreement with known experimental data. The equations must reduce to the Lorentz-Poincaré transformations when $v_{01} = v_{02}$, and they must explain why the experimental measurement of the speed of light by the usual methods always gives C.

We first note that when $v_{01} = v_{02}$ the frames S_1 and S_2, which coincided at the initial instant, always remain coincident. This brings us back to the case of two inertial frames, where one of them is at rest in the Cosmic Substratum (aether frame). In this case, as expected, $\Delta = v_{01} \ell_0 / C^2$, which is the synchronism discrepancy effect defined by Prokhovnik.[22] (Actually, Prokhovnik takes $\delta = \gamma(v_{01} \ell_0 / C^2)$ as a definition of the concept. But, contrary to this author's opinion, this implies that the measurement is carried out with non-retarded clocks. See Chapter 6.)

We can see that in this case

$$X_{1app} = \ell_0,$$

and

$$X_{1r} = \ell_0 \sqrt{1 - v_{01}^2 / C^2} = X_{1app} \sqrt{1 - v_{01}^2 / C^2}.$$

Referring to the co-ordinates of frame S_0 by the letters X and T, we remark that

$$X = X_{1r} + v_{01} T$$
$$= X_{1app} \sqrt{1 - v_{01}^2 / C^2} + v_{01} T,$$

so

$$X_{1app} = \frac{X - v_{01} T}{\sqrt{1 - v_{01}^2 / C^2}}. \qquad (4.10)$$

Now, taking the synchronism discrepancy Δ of two clocks placed at the ends of the rod into account, we have:

$$T_{1app} = T\sqrt{1-v_{01}^2/C^2} - v_{01}\frac{X_{1app}}{C^2}.$$

Replacing T by its value

$$\frac{\ell_0\sqrt{1-v_{01}^2/C^2}}{V-v_{01}}, \tag{4.11}$$

we obtain:

$$\begin{aligned}T_{1app} &= \frac{X_{1app}(1-v_{01}^2/C^2)}{V-v_{01}} - \frac{v_{01}}{C^2}X_{1app}\\ &= X_{1app}\frac{(C^2-v_{01}V)}{C^2(V-v_{01})}.\end{aligned}$$

From expression (4.11) we obtain

$$\begin{aligned}T_{1app} &= \frac{T(C^2-v_{01}V)}{C^2\sqrt{1-v_{01}^2/C^2}}\\ &= \frac{T-v_{01}X/C^2}{\sqrt{1-v_{01}^2/C^2}}.\end{aligned} \tag{4.12}$$

From (4.10) and (4.12) we deduce the reciprocal co-ordinates

$$X = \frac{X_{1app} + v_{01}T_{1app}}{\sqrt{1-v_{01}^2/C^2}},$$

and

$$T = \frac{T_{1app} + v_{01}X_{1app}/C^2}{\sqrt{1-v_{01}^2/C^2}}.$$

Thus, as expected, when $v_{01} = v_{02}$ the space-time transformations connecting the frames S_0 and S_1 reduce to the Lorentz-Poincaré transformations.

For $V = C$, the apparent time and space co-ordinates reduce to:

$$T_{1app} = \frac{\ell_0}{C}\frac{\sqrt{1-v_{02}^2/C^2}}{\sqrt{1-v_{01}^2/C^2}}\frac{C-v_{01}}{C-v_{02}},$$

and

$$X_{1app} = \ell_0\frac{\sqrt{1-v_{02}^2/C^2}}{\sqrt{1-v_{01}^2/C^2}}\frac{C-v_{01}}{C-v_{02}}.$$

It can be seen that the *apparent* (measured) speed of light V_{1app} in frame S_1 is equal to C, in conformity with experiment.

Now, when $v_{01} = 0$, S_1 is at rest in the Cosmic Substratum,

$$X = \ell_0 \frac{\sqrt{1 - v_{02}^2/C^2}}{V - v_{02}} V,$$

and

$$T = \ell_0 \frac{\sqrt{1 - v_{02}^2/C^2}}{V - v_{02}},$$

$$T = \frac{\ell_0 \sqrt{1 - v_{02}^2/C^2}}{C - v_{02}} \frac{C - v_{02}}{V - v_{02}}$$

$$= \frac{\ell_0/C + v_{02}\ell_0/C^2}{\sqrt{1 - v_{02}^2/C^2}} \frac{C - v_{02}}{V - v_{02}}.$$

After multiplication of the two fractions we obtain

$$T = \frac{\ell_0 - v_{02}^2 \ell_0/C^2}{\sqrt{1 - v_{02}^2/C^2} \ (V - v_{02})}$$

$$= \frac{\ell_0(1 - v_{02}V/C^2) + (V - v_{02})v_{02}\ell_0/C^2}{\sqrt{1 - v_{02}^2/C^2} \ (V - v_{02})}.$$

When allowance is made for the fact that

$$V_{2app} = \frac{V - v_{02}}{1 - v_{02}V/C^2} \text{ and } X_{2app} = \ell_0,$$

the expression for T becomes

$$T = \frac{T_{2app} + v_{02}X_{2app}/C^2}{\sqrt{1 - v_{02}^2/C^2}}. \tag{4.13}$$

Multiplying T by V so that:

$$X = \frac{T_{2app} + v_{02}X_{2app}/C^2}{\sqrt{1 - v_{02}^2/C^2}} \times V, \tag{4.14}$$

and making allowance for the expression

$$V = \frac{V_{2app} + v_{02}}{1 + v_{02}V_{2app}/C^2},$$

we easily find from (4.14) that:

$$X = \frac{X_{2app} + v_{02}T_{2app}}{\sqrt{1 - v_{02}^2/C^2}}.$$

Therefore, as expected, we obtain the Lorentz-Poincaré transformations when one of the frames is at rest in the Cosmic Substratum.

Important remark

ℓ_0 is not the real co-ordinate of point A relative to S_2 along the x_2-axis; the real coordinate is ℓ. The fact that authors are not aware of this is a source of much confusion. It should also be pointed out that, contrary to what is often believed, X_{1app}, T_{1app} and V_{1app} are all apparent (fictitious) co-ordinates.

Conclusion

Starting from the Galilean transformations and assuming the Lorentz postulates, we have obtained a set of transformations applicable to all pairs of inertial bodies aligned along the direction of motion of the Solar system, even if none of the bodies is at rest in the Cosmic Substratum (aether frame). These relations assume a form different from the Lorentz-Poincaré transformations.

In order to derive them, we were compelled to modify the Galilean transformations by making allowance for the systematic errors of measurement. The extended space-time transformations obtained in this way are the experimental relations. Conversely, these transformations must be corrected in order to obtain the Galilean relations, which are the true transformations when no errors of measurement are present.

The derivation is demonstrated to be consistent, since the extended transformations reduce to the Lorentz-Poincaré transformations when one of the frames under consideration is the fundamental inertial frame. They also explain why the *'apparent'* (measured) velocity of light is found to be constant independently of the absolute speed of the Solar system. However, after correction of the systematic errors of measurement, they also show that the real velocity of light is constant only in the aether frame.

These extended space-time transformations do contradict the relativity principle. Yet this does not disprove them since, as we saw at the end of Chapter 2 and as we will see in the following text, while the principle can be used as a good approximation in different commonly encountered cases,[5] it cannot be generalized to all experimental situations.

It is worth noting that the transformations derived here are of more than theoretical interest. They reflect our actual situation in the Cosmos since, as we have seen in the previous chapters, the solar system is in motion with respect to the Cosmic Substratum with an estimated velocity of about 300 km/sec.[15,30,31]

Another important point needs to be clarified. We know that in aether theories, contrary to relativity, the kinetic energy presents an absolute character. It is defined with respect to the fundamental frame S_0. This means that the increase of a body's kinetic energy when it passes from one inertial frame S_1 (distinct from S_0) to another S_2, is different from the conventional value. The calculation is as follows:

When the body passes from S_0 to S_1, the kinetic energy acquired is

$$(m_1 - m_0) C^2.$$

If the speed of the body is low ($v_{01}/C \ll 1$), this expression reduces to

$$\frac{1}{2} m_0 v_{01}^2.$$

When the body passes from S_0 to S_2 (with $v_{02}/C \ll 1$), the kinetic energy acquired is

$$\frac{1}{2} m_0 v_{02}^2.$$

The increase of kinetic energy from S_1 to S_2 is thus

$$\frac{1}{2} m_0 \left(v_{02}^2 - v_{01}^2\right). \tag{4.15}$$

Since $v_{02} \ll C$, we can write, $v_{12app} \cong v_{12} = v_{02} - v_{01}$. Expression (4.15) then becomes:

$$\frac{1}{2} m_0 \left(v_{12}^2 + 2 v_{01} v_{12}\right). \tag{4.16}$$

This expression differs from the conventional formula, i.e.,

$$\frac{1}{2} m_0 v_{12}^2.$$

Indeed, unlike this expression, formula (4.16) depends on v_{01}, which is the speed of the Earth with respect to the fundamental frame. In earlier work[3] we viewed this result as an argument against aether theories, but in light of new data that refute the relativity principle,[5,6] we think that our previous position deserves further consideration.

When we analyze expression (4.16), we note that when $v_{12} \gg v_{01}$ the term depending on v_{01} becomes relatively insignificant and we are brought back to the conventional expression. In all other cases, v_{01} cannot be ignored, since it is estimated at about 300 km/sec.

It should be noted that, in conventional physics, when the relativity principle is assumed, the expression for the kinetic energy is plagued by a serious internal contradiction. Indeed, when a body passes from one inertial system S_0 to another S_1, assuming that $v_{01} \ll C$, it acquires the kinetic energy.

$$(m_1 - m_0) C^2 \cong \frac{1}{2} m_0 v_{01}^2. \qquad (4.17)$$

Now, when the same body passes from S_0 to S_2 with $v_{02} \ll C$, the kinetic energy acquired is

$$(m_2 - m_0) C^2 \cong \frac{1}{2} m_0 v_{02}^2. \qquad (4.18)$$

The difference between (4.18) and (4.17) is

$$\frac{1}{2} m_0 \left(v_{02}^2 - v_{01}^2 \right),$$

which is different from $m_0 v_{12}^2 / 2$. (Note that, contrary to Lorentz aether theories, v_{01} is not the speed of the body with respect to a preferred frame.)

Suppose that $v_{01} = v_{12} = v$. Assuming that $v \ll C$, we easily check that the kinetic energy increase from S_1 to S_2 is $3 m_0 v^2 / 2$. But according to the relativity principle it should be $m_0 v^2 / 2$, since nothing differentiates the three inertial frames. This important internal contradiction does not affect the fundamental aether theory.

The paradox appears even more obvious if we note that the energy acquired when a body passes from one inertial frame S_0 to S_1, is not clearly defined and depends on the frame from which it is measured. Suppose that $v_{01} = 0$, and that S_0 and S_1 coincide. If $v_{12} = 1$ km/sec, the kinetic energy acquired from S_1 to S_2 will be equal to $m_0 / 2$. If S_1 moves with respect to S_0 at 10 km/sec, even though v_{12} has not changed, the kinetic energy needed to pass from S_1 to S_2, measured from frame S_0, will appear to be:

$$\frac{1}{2} m_0 \left(11^2 - 10^2 \right) = \frac{1}{2} m_0 (121 - 100) = 10.5 \, m_0. \qquad (4.19)$$

This is in contradiction with the idea that the energy needed to perform a certain amount of work is well defined and cannot depend on the point from which it is measured.

This paradox is completely foreign to the fundamental aether theory, where the energy is perfectly defined and depends on the speed of the body with respect to the aether frame.

Appendix 1

Let us consider the case of a light signal which starts from the common origin and propagates toward point A (Figure 7.) When the signal reaches point A, it is reflected in a mirror firmly attached to frame S_1 at a point A' in the frame which at this instant coincides with A. After reflection, the signal comes back to O'.

1 - Real co-ordinates

The ratio of the distances covered by the signal from O' to A' and from O'' to A is equal to the ratio of the speeds of the signal with respect to S_1 and to S_2.

$$\frac{x_1}{\ell_0\sqrt{1-v_{02}^2/C^2}} = \frac{C-v_{01}}{C-v_{02}},$$

and hence

$$x_1 = \ell_0\sqrt{1-v_{02}^2/C^2}\,\frac{C-v_{01}}{C-v_{02}}.$$

Therefore, the real time needed by the signal to cover the distance x_1 is:

$$t_1 = \frac{x_1}{C-v_{01}} = \frac{\ell_0\sqrt{1-v_{02}^2/C^2}}{C-v_{02}}, \qquad (4.20)$$

where $C-v_{01}$ is the real velocity of light in frame S_1, and $C-v_{02}$ the real velocity in frame S_2. It should be borne in mind that these lengths, times and speeds are not what are measured experimentally.

2 - Apparent co-ordinates

The experimental (*apparent*) time can easily be obtained from the real time by allowing for the systematic errors of measurement. This is also true of the apparent path. Since the distance x_1 is measured with a contracted meter stick, it appears longer than it really is, and the apparent distance is then:

$$x_{1app} = \frac{x_1}{\sqrt{1-v_{01}^2/C^2}} = \frac{\ell_0\sqrt{1-v_{02}^2/C^2}}{\sqrt{1-v_{01}^2/C^2}}\,\frac{C-v_{01}}{C-v_{02}}.$$

As we saw above, the time needed by the light signal to go from O' to A' is erroneously identified with the '*apparent*' average transit time, which is equal to:

$$\frac{t_1+\bar{t}_1}{2}\sqrt{1-v_{01}^2/C^2}, \qquad (4.21)$$

where t_1 is given by formula (4.20).

We can see that in the reverse direction (A'→O'), the light signal covers, with respect to S_1, the same distance as from O' to A', but with the speed $C+v_{01}$.

Thus,

$$\bar{t}_1 = \frac{x_1}{C+v_{01}} = \ell_0\sqrt{1-v_{02}^2/C^2}\,\frac{C-v_{01}}{C-v_{02}} \times \frac{1}{C+v_{01}},$$

and

$$\frac{t_1 + \bar{t}_1}{2} = \frac{1}{2} \ell_0 \sqrt{1 - v_{02}^2/C^2} \frac{C - v_{01}}{C - v_{02}} \left(\frac{1}{C - v_{01}} + \frac{1}{C + v_{01}} \right).$$

When this result is inserted in (4.21), the *apparent* average transit time of light in frame S_1, d_{1app}, reduces to:

$$\ell_0 C \frac{\sqrt{(1 - v_{01}^2/C^2)(1 - v_{02}^2/C^2)}}{(C + v_{01})(C - v_{02})}. \tag{4.22}$$

After simplification expression (4.22) becomes:

$$d_{1app} = \frac{\ell_0}{C} \frac{\sqrt{1 - v_{02}^2/C^2}}{\sqrt{1 - v_{01}^2/C^2}} \frac{C - v_{01}}{C - v_{02}}. \tag{4.23}$$

3 - Apparent speed of light

The apparent (experimental) speed of light, as expected, emerges as:

$$\frac{x_{1app}}{d_{1app}} = C.$$

N.B. The slow clock transport method is affected by similar systematic errors of measurement.

Appendix 2

Is the existence of a preferred reference frame compatible with group structure?

Poincaré made a vital contribution to the relativistic hypothesis by demonstrating that the relativity principle necessarily implies a group structure for the space-time transformations. What this means is that, if we assume the equivalence of all inertial frames for the description of physical laws, then the space-time transformations between any pair of such reference systems must be of identical mathematical form.

The exact demonstration of this statement will be recalled at the end of the present Appendix. Nevertheless, the following question is worth asking: are all inertial frames actually equivalent? According to Einstein, there is no doubt about this, but, if we assume the Lorentz postulates, this is not so obvious. In fact, in Lorentz's view:

1. There exists a preferred reference frame attached to the aether (fundamental inertial frame).
2. The contraction of moving rods is real and non-reciprocal.
3. The one-way speed of light is isotropic and equal to C only in the fundamental inertial frame (Cosmic Substratum).

Starting from these same hypotheses, in Appendix 1 we studied the example of a light signal travelling along the *x*-axis, and demonstrated that the transformations for space between any pair of inertial systems S_1 and S_2 take, in this case, the following form

$$x_1 = \ell_0 \sqrt{1 - v_{02}^2/C^2} \, \frac{C - v_{01}}{C - v_{02}}.$$

Because distance x_1 is measured with a contracted meter stick in frame S_1, its apparent length becomes:

$$x_{1app} = \ell_0 \frac{\sqrt{1 - v_{02}^2/C^2}}{\sqrt{1 - v_{01}^2/C^2}} \frac{C - v_{01}}{C - v_{02}}. \tag{4.24}$$

We see that when S_1 is at rest in the Cosmic Substratum S_0, then $v_{01} = 0$ and expression (4.24) reduces to:

$$x_1 = x_{2app} \sqrt{1 - v_{02}^2/C^2} \, \frac{C}{C - v_{02}}.$$

(Note that ℓ_0 is the length of the rod when it is at rest in frame S_0, but it is also the apparent length x_{2app} of the rod in motion as measured by observer S_2 with his contracted meter stick.)

Now since $v_{02} = v_{12}$

$$x_1 = \frac{x_{2app} + v_{12} x_{2app}/C}{\sqrt{1 - v_{12}^2/C^2}}. \tag{4.25}$$

Replacing x_{2app} by its value Ct_{2app} in (4.25), we obtain the classical Lorentz-Poincaré transformation.

$$x_1 = \frac{x_{2app} + v_{12} t_{2app}}{\sqrt{1 - v_{12}^2/C^2}}. \tag{4.26}$$

Thus, the classical transformation applies only when one of the frames is at rest in the Cosmic Substratum.

Transformation (4.24) is different from Einstein's relativistic transformation, since it depends on v_{01} and v_{02}. But v_{01} and v_{02} are the speeds of reference frames S_1 and S_2 with respect to the fundamental frame S_0 which, as we can see, is omnipresent when one assumes the Lorentz postulates. Einstein's relativistic approach is completely different. Here v_{01} and v_{02} do not mean anything, and C is the real speed of light. Einstein's transformation between frames S_1 and S_2 is:

$$x_{1E} = \frac{x_{2E} + v_{12} t_{2E}}{\sqrt{1 - v_{12}^2/C^2}}$$

(E = Einstein). It is different from (4.24), but assumes a form similar to (4.26). In Einstein's theory, it is valid for any pair of inertial frames.

The same remark applies to the transformations governing the clock displays. (See formula (4.23).)

$$d_{1app} = \frac{\ell_0}{C}\frac{\sqrt{1-v_{02}^2/C^2}}{\sqrt{1-v_{01}^2/C^2}}\frac{C-v_{01}}{C-v_{02}} = \frac{\ell_0}{C}\sqrt{\frac{C+v_{12}-v_{01}v_{02}/C}{C-v_{12}-v_{01}v_{02}/C}}, \qquad (4.27)$$

but
$$t_{1E} = \frac{\ell_0/C+v_{12}\ell_0/C^2}{\sqrt{1-v_{12}^2/C^2}} = \frac{\ell_0}{C}\sqrt{\frac{C+v_{12}}{C-v_{12}}}. \qquad (4.28)$$

Expressions (4.27) and (4.28) differ in the terms depending on v_{01} and v_{02}. These terms are negligible when v_{01} and $v_{02} \ll C$, but become more and more significant as v_{01} and v_{02} increase.

Notice that in the present example of a light signal: $x_{1app}/d_{1app} = C$. But contrary to Einstein's special relativity, C is not the real speed of light in S_1: it is the *'apparent'* average two way-speed of light measured with retarded clocks and contracted meter sticks. Note also that the space-time transformations derived from the Lorentz postulates do not assume a group structure. In summary:

$S_0 \xrightarrow{T_{01}} S_1$ is a classical Lorentz-Poincaré transformation;
$S_0 \xrightarrow{T_{02}} S_2$ is a classical Lorentz-Poincaré transformation;
but $S_1 \xrightarrow{T_{12}} S_2$ is not.

Therefore, contrary to special relativity: $T_{01}T_{12} \neq T_{02}$. This seems paradoxical, but is easily explained when we know that in frame S_0, measurements of distances and time are exact, whereas those carried out in frame S_1 and S_2 are fictitious. Moreover in frame S_0 the speed of light is isotropic, but not in S_1 and S_2.

Poincaré was persuaded of the interest presented by a relativity principle and the group structure associated with it, but, at the same time, he believed in the postulates of Lorentz's theory and, although these concepts seemed to be in conflict, he thought that they could be reconciled, at least apparently. He did not deny the existence of a particular inertial frame attached to the aether, the real and non-reciprocal contraction of moving lengths, and the necessity of a medium to convey electromagnetic waves.

On page 1 of his article "*Sur la dynamique de l'électron*,"[18] he expressed the relativity principle as follows:

> It appears that the impossibility *of observing the absolute motion of Earth* is a general law of nature. We are naturally led to assume this law, which we will refer to as the relativity postulate.

At the end of section 7, speaking of the Fitzgerald-Lorentz contraction (real and non reciprocal), he states:

> ....Therefore, the hypothesis of Lorentz (contraction) is the only one which is compatible with the impossibility of bringing absolute motion into evidence. (See also the quotations from Poincaré given in the introduction of the present book.)

It is thus clear that for Poincaré absolute motion exists. Moreover, Lorentz contraction implies absolute motion, since, if a rod really contracts when it moves from one inertial frame to another, it is because there is a hierarchy between the different inertial frames (and not an equivalence). Therefore Poincaré's theory sought to reconcile two incompatible notions: the relativity principle on the one hand, and the existence of an absolute aether frame on the other.

We can therefore say that group structure applies to Einstein's transformations, but not to the transformations derived from the Lorentz postulates (including expressions (4.24) and (4.27)). (Note nevertheless that Poincaré stated many times that he did not acknowledge Newton's absolute space.)

Note: The Lorentz group according to Poincaré

In section 4 of his article "*Sur la dynamique de l'électron*,"[18] Poincaré demonstrates that the equivalence of all inertial frames (relativity principle) implies a group structure. There is nothing objectionable about this if we suppose that all frames are actually equivalent, but, as we have seen, this is not the case if we adopt Lorentz's postulates. (See formulae (4.24) and (4.27).)

This can be seen in the following comments regarding the demonstration of the space transformations, given in outline by Poincaré and expounded here in more detail. Let us consider three frames S″, S′ and S. (See Figure 8.) If we suppose that they are equivalent, we have:

$$x' = \frac{x+vt}{\sqrt{1-v^2/C^2}}$$

$$t' = \frac{t+vx/C^2}{\sqrt{1-v^2/C^2}}$$

$$x'' = \frac{x'+v't'}{\sqrt{1-v'^2/C^2}}$$

$$t'' = \frac{t'+v'x'/C^2}{\sqrt{1-v'^2/C^2}}.$$

(4.29)

Here we have adopted the same notation as Poincaré did for S, S′ and S″. Replacing x' and t' by their values in x'' we obtain:

$$x'' = \frac{N}{D} = \frac{x+vt+v'(t+vx/C^2)}{\sqrt{(1-v^2/C^2)(1-v'^2/C^2)}}.$$

The numerator N can be written as follows:

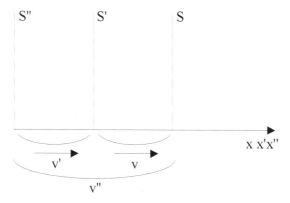

Figure 8 - The Lorentz group according to Poincaré.

$$N = x + \frac{(v+v')t}{1+vv'/C^2}\left(1+vv'/C^2\right) + vv'x/C^2 .$$

Let us define

$$v'' = \frac{v+v'}{1+vv'/C^2} .$$

We then obtain:

$$N = x + v''t + \frac{vv'}{C^2}(x+v''t) = (x+v''t)\left(1+\frac{vv'}{C^2}\right).$$

Therefore,

$$x'' = \frac{(x+v''t)(1+vv'/C^2)}{\sqrt{(1-v^2/C^2)(1-v'^2/C^2)}} . \tag{4.30}$$

We easily demonstrate that expression (4.30) reduces to:

$$x'' = \frac{x+v''t}{\sqrt{1-v''^2/C^2}} .$$

Therefore, the equivalence of all inertial frames implies a group structure.

But, as demonstrated previously, if S″ is the fundamental frame, the space-time transformations $x' = f(x,t)$ and $t' = f(x,t)$ in (4.29) are not compatible with the Lorentz postulates; only expressions (4.7) and (4.8) are. As a result the space-time transformations do not constitute a group.

Chapter 5

Inertial Transformations Derived from Galilean Transformations

Prologue

This chapter deals with the "inertial transformations" as defined by Franco Selleri. These transformations are equivalent to the extended space-time transformations, although they assume a different mathematical form. They presuppose an exact synchronization of clocks in the Earth frame, which is ideal but not usual and would be difficult to carry out.

This chapter is independent of the following chapter, so the reader may skip it in a first reading. However, it does contain important information and should not be ignored.

We are indebted to Builder and Prokhovnik[22] for demonstrating that the term vx'/C^2 of the Lorentz-Poincaré transformation for time is nothing more than an artefact that arises from the synchronization procedure, *i.e.*, Einstein-Poincaré synchronization. Tangherlini[32] (1961) and Mansouri and Sexl[29] (1977), meanwhile, proposed a set of transformations that do not suffer from the "synchronism discrepancy effect." Their transformations assume the following form:

$$t = t_0\sqrt{1-v^2/C^2},$$

$$x = \frac{x_0 - vt_0}{\sqrt{1-v^2/C^2}}.$$

According to Mansouri and Sexl, x_0 and t_0 are the co-ordinates of any inertial frame taken as a reference system, and x and t the co-ordinates of any other inertial frame. In fact, since this set of transformations is equivalent to the Lorentz-Poincaré transformations, it must apply exactly only when x_0 and t_0 are the co-ordinates of the fundamental inertial frame.

40   Joseph Lévy

The transformations were extended to all inertial frames by Franco Selleri[28] from the following assumptions: existence of an aether frame in a state of absolute rest, isotropy of the experimental two-way velocity of light, isotropy and homogeneity of space and time, time dilation, and some other simplifying constraints.

We propose here to demonstrate that these inertial transformations can be obtained in a simpler manner from the Galilean transformations, by introducing the systematic errors due to length contraction and clock retardation. We will first derive the inertial transformations between the aether frame and other inertial frames, and then the extended inertial transformations.

Inertial transformations between the aether frame and other inertial frames

1 - Space transformations

Consider a system of co-ordinates S_0 (x_0, y_0, z_0, t_0) at rest with respect to the Cosmic Substratum (aether frame) and a long rod, O'x, moving along the x_0 axis with rectilinear uniform motion. The rod is firmly attached to an inertial frame S (x, y, z, t) receding from S_0 at speed v_0.

At the initial instant (0), O and O' are coincident. At this very instant, two inertial bodies 1 and 2 coming from the $-x_0$ direction and running along the rod in the $+x_0$ direction, pass, side by side, at point O, O'.

At time t_0 (measured with a clock attached to frame S_0 placed at point A) body 1 has covered the distance O'A = ℓ_1 in reference frame S. At the same instant t_0 (measured with a clock attached to frame S_0 placed at B), body 2 has covered the distance O'B = ℓ_2.

Notice that there is no ambiguity for the time, since clocks A and B are at rest with respect to the aether frame where the speed of light is isotropic and there are no theoretical difficulties in synchronizing them exactly. (See Figure 9; of course this a thought experiment.)

The general expression of the Galilean transformations for distances is

$$x_0 = v_0 t_0 + x. \qquad (5.1)$$

Therefore, the space coordinate of point A is:

$$x_{01} = v_0 t_0 + \ell_1, \qquad (5.2)$$

and the coordinate of point B is:

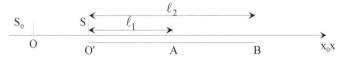

Figure 9. At the instant t_0, body 1 has covered the distance ℓ_1, while body 2 has covered the distance ℓ_2.

$$x_{02} = v_0 t_0 + \ell_2, \tag{5.3}$$

so that

$$x_{02} - x_{01} = \ell_2 - \ell_1.$$

ℓ_1 and ℓ_2 are the real distances covered by the bodies. The lengths ℓ_1 and ℓ_2 are contracted, but the contraction is not apparent in expressions (5.2) and (5.3). Now according to the Lorentz assumptions, since these distances are measured with a contracted meter stick in reference frame S, they appear to have different lengths L_{1app} and L_{2app} such that:

$$L_{1app} = \frac{\ell_1}{\sqrt{1 - v_0^2/C^2}}, \tag{5.4}$$

and

$$L_{2app} = \frac{\ell_2}{\sqrt{1 - v_0^2/C^2}}, \tag{5.5}$$

where the suffix "$_{app}$" indicates that the result of these measurements is apparent and not real.

Note that L_{1app} and L_{2app} are the real lengths that O'A and O'B would have if they were at rest in frame S_0. (It should be pointed out that according to the Lorentz assumptions, it is matter that is contracted, not space).

For any distance x we would have

$$X_{app} = \frac{x}{\sqrt{1 - v_0^2/C^2}}. \tag{5.6}$$

Substituting (5.6) into (5.1) we obtain

$$x_0 = v_0 t_0 + X_{app} \sqrt{1 - v_0^2/C^2}.$$

Hence

$$X_{app} = \frac{x_0 - v_0 t_0}{\sqrt{1 - v_0^2/C^2}}. \tag{5.7}$$

Expressions (5.1) and (5.7) are equivalent. So there is no incompatibility between the Lorentz-Poincaré and the Galilean transformations regarding space. The former connects the real distance $x_0 - v_0 t_0$ covered by a body in frame S, with the apparent (fictitious) distance X_{app} measured with a contracted meter stick in this frame; while the Galilean transformation connects $x_0 - v_0 t_0$ with the real distance x covered by the body (obtained after correction of the systematic error of measurement).

Substituting (5.4) and (5.5) in (5.2) and (5.3) we obtain:

$$L_{1app} = \frac{x_{01} - v_0 t_0}{\sqrt{1 - v_0^2/C^2}},$$

$$L_{2app} = \frac{x_{02} - v_0 t_0}{\sqrt{1 - v_0^2/C^2}},$$

so that

$$(L_2 - L_1)_{app} = \frac{x_{02} - x_{01}}{\sqrt{1 - v_0^2/C^2}},$$

where $x_{02} - x_{01}$ is the real length of the contracted distance AB measured with a non-contracted meter stick from frame S_0, and $(L_2 - L_1)_{app}$ is the fictitious length of AB measured with a contracted meter stick in frame S.

This demonstrates that the relative distances and speeds of moving bodies measured by an observer at rest on a planet are affected by systematic errors which require correction. The errors could be substantial if the planet is moving at high speed with respect to the Cosmic Substratum. It is also important to note that the Lorentz-Fitzgerald contraction appears compatible with the Galilean transformations.

2 - Time transformations

Tangherlini[32] and Mansouri and Sexl[29] both noticed that, assuming the postulates of Lorentz, and with a suitable choice of clock synchronization, the relation between the clock displays of reference frames S_0 (x_0, y_0, z_0, t_0) and S (x, y, z, t) is given by

$$t = t_0 \sqrt{1 - v_0^2/C^2} \ . \tag{5.8}$$

Indeed, from formula (4.13) for example, the Lorentz-Poincaré transformation for time can be written as follows:

$$t_0 = \frac{T_{app} + v_0 X_{app}/C^2}{\sqrt{1 - v_0^2/C^2}},$$

so that

$$T_{app} = t_0 \sqrt{1 - v_0^2/C^2} - v_0 X_{app}/C^2 \ . \tag{5.9}$$

From Chapter 3 of the present volume, we can infer that the $v_0 X_{app}/C^2$ term, results from the error introduced when we synchronize clocks with light rays, whose speed is assumed to be isotropic in the Earth frame. This term is only conventional and can be eliminated.[1,2,22] Therefore, with a suitable choice of clock synchronization, (5.9) reduces to (5.8).

3 - Law of composition of velocities

In the Galilean view, the law of composition of velocities takes the form

$$V = v_0 + v_r, \tag{5.10}$$

where V is the speed of a body with respect to the aether frame S_0, v_0 is the speed of reference frame S with respect to S_0 and v_r is the real speed of the body relative to S. We must specify that all these measurements are carried out by an observer at rest in frame S_0 and that the body moves along the x_0, x-axis at uniform speed.

For a body that has covered the real distance ℓ in S during the real time t_0:

$$v_r = \frac{\ell}{t_0},$$

but when observer S measures the distance ℓ with his contracted meter stick, he finds:

$$L_{app} = \frac{\ell}{\sqrt{1-v_0^2/C^2}},$$

as for the clock display obtained with the retarded clock of frame S (which we shall call t_{app} in order to distinguish it from T_{app}):

$$t_{app} = t_0\sqrt{1-v_0^2/C^2}.$$

Consequently, the *apparent* (fictitious) speed measured by observer S is:

$$v_{app} = \frac{L_{app}}{t_{app}} = \frac{v_r}{1-v_0^2/C^2}. \tag{5.11}$$

Therefore, from (5.10) and (5.11) the law of composition of velocities becomes:

$$V = v_0 + v_{app}(1-v_0^2/C^2), \tag{5.12}$$

and

$$v_{app} = \frac{V-v_0}{1-v_0^2/C^2}$$

This law has the same form as the law obtained by a different route by Mansouri and Sexl; but we now realize that v_{app} is not the real speed. Note also that (5.12) is equivalent to (5.10) and is nothing more than the Galilean law of composition of velocities obtained when we use contracted meter sticks and retarded clocks. (Since the speeds have been measured with exactly synchronized clocks, its form is different from the Lorentz-Poincaré law.)

Note that v_0 is the real speed of frame S with respect to S_0. If the speed had been measured by observer S with his contracted meter stick and retarded clock, the apparent value of this speed would have been:

$$v_{0app} = \frac{OO'}{\sqrt{1-v_0^2/C^2}} \bigg/ t_0\sqrt{1-v_0^2/C^2} = \frac{v_0}{1-v_0^2/C^2}.$$

Compatibility of the Galilean law of composition of velocities with a limit velocity

If we suppose that v_0 and v_r can take any value between 0 and C, there is no possible compatibility between the Galilean law of composition of velocities and the existence of a limit velocity. The conventional belief is that v_r is not dependent on v_0. For this reason, in previous papers, we were convinced that the Galilean theory, which is supposed to allow speeds higher than C, could not be in agreement with the law $m = m_0 \gamma$ (variation of mass with speed.)

But this apparent obstacle is due to erroneous suppositions. Actually, the relationship

$$V = v_0 + v_r < C$$

imposes a limit on v_r such that

$$v_r < C - v_0,$$

and consequently,

$$v_{app} < \frac{C - v_0}{1 - v_0^2/C^2}.$$

So, contrary to conventional belief, the law $m = m_0 \gamma$ is not incompatible with the Galilean law of composition of velocities.

Inertial transformations in the general case

We will now derive the inertial transformations connecting two inertial frames S_1 and S_2 moving with respect to the aether frame S_0 at speeds v_{01} and v_{02} along the x_0-axis (Figure 10).

At the initial instant (0), O, O' and O'' are coincident. At this very instant a body coming from the left, passes by the common origin, and then continues on its way with rectilinear uniform motion toward a point A. The body moves along a long rod O''x_2 which is at rest with respect to the S_2 frame.

At time t_0 (for S_0), the body has covered a distance ℓ in S_2, which is equal to the real length of O''A measured from reference frame S_0, with a non-contracted meter stick.

We will use the subscript "$_r$" for real and "$_{app}$" for apparent. The Galilean relationship for $x_{1r} = $ O'A gives:

$$x_{1r} = (v_{02} - v_{01})t_0 + \ell, \qquad (5.13)$$

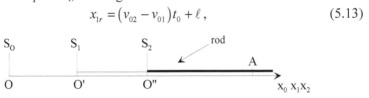

Figure 10. At instant t_0 the body has covered the distance O''A in reference frame S_2.

where v_{01} and v_{02} are, respectively, the real speeds of S_1 and S_2 with respect to S_0.

According to length contraction

$$\ell = \ell_0 \sqrt{1 - v_{02}^2/C^2} , \qquad (5.14)$$

where ℓ_0 is the length of O''A when it is at rest in frame S_0. Note that $(v_{02} - v_{01})t_0$ is not contracted because it is not a rigid object moving with respect to the aether frame. It is only the space separating the two reference frames S_1 and S_2.

Now, using a contracted meter stick, observer S_1 will find an apparent distance x_{1app} for O'A such that (from (5.13) and (5.14)):

$$x_{1app} = \frac{(v_{02} - v_{01})t_0}{\sqrt{1 - v_{01}^2/C^2}} + \frac{\ell_0 \sqrt{1 - v_{02}^2/C^2}}{\sqrt{1 - v_{01}^2/C^2}} . \qquad (5.15)$$

For observer S_2, O''A does not appear modified, since the meter stick he uses is also contracted (in the same ratio as O''A.) So:

$$\ell_0 = x_{2app} . \qquad (5.16)$$

Now observer S_1 makes use of a clock whose display t_{1app} is different from the real time in such a way that:

$$t_{1app} = t_0 \sqrt{1 - v_{01}^2/C^2} . \qquad (5.17)$$

Utilizing (5.16) and (5.17), expression (5.15) can be written as follows:

$$x_{1app} = \frac{(v_{02} - v_{01})t_{1app}}{1 - v_{01}^2/C^2} + x_{2app} \frac{\sqrt{1 - v_{02}^2/C^2}}{\sqrt{1 - v_{01}^2/C^2}} . \qquad (5.18)$$

From expression (5.18) we easily obtain x_{2app}:

$$x_{2app} = \left(x_{1app} - \frac{(v_{02} - v_{01})t_{1app}}{1 - v_{01}^2/C^2} \right) \frac{\sqrt{1 - v_{01}^2/C^2}}{\sqrt{1 - v_{02}^2/C^2}} . \qquad (5.19)$$

Expression (5.19) is identical in form to the transformation obtained by Franco Selleri[28] (p 346, formula (34)). However, here we have established that it is actually a relationship between *apparent* times and lengths.

For the clock display we easily obtain:

$$t_{2app} = t_{1app} \frac{\sqrt{1 - v_{02}^2/C^2}}{\sqrt{1 - v_{01}^2/C^2}} .$$

N.B. Length contraction changes our evaluation of x_1 so that x_{1r} becomes x_{1app}. But the changed clock display has no influence on x_1.

Chapter 6

Arguments in Favour of Lorentz-Fitzgerald Contraction

Introduction

Since the early days of relativity theory, Lorentz-Fitzgerald contraction has been the focus of a debate which is still lively today, and divides physicists into opposing camps. Some—such as Wesley,[31] Phipps,[33] Cornille,[34] Galeczki[35]—regard length contraction as a naïve conjecture.

Others consider it a fundamental process that explains a vast body of experimental facts; among them are Bell,[36] Selleri,[37] Builder, Prokhovnik,[22] Dishington,[38] Mansouri and Sexl,[29] and Wilhelm.[30]

Length contraction was originally proposed by Lorentz and Fitzgerald in order to explain the null result of the Michelson experiment. (In fact, the result was not completely null, but much smaller than expected.) Length contraction has never been observed: of course, it cannot be demonstrated by an observer of the moving frame, since the standard used to measure it also contracts. But it might be verified indirectly. This was the objective of various renowned physicists who tried to observe the physical changes brought about by motion: variation of the refractive index of a refringent solid (Rayleigh[39] and Brace[40]); influence of the aether wind on a charged condenser whose plates form a certain angle with the direction of translation (Trouton and Noble[41]); experiments of Trouton and Rankine[42] and of Chase[43] and Tomashek[44] on the electrical resistance of moving objects; and finally the Wood, Tomlinson and Essex[45] experiments on the frequency of the longitudinal vibration of a rod. Yet the experiments all proved negative.

Nevertheless, Lorentz was able to explain the negative result by assuming an increase of mass with speed.[46] At first sight Lorentz's argument appeared dubious. But, as we saw in Chapter 4, once the relativity principle is suspended, the objections no longer hold. In Chapter 8, we will give a conclusive demonstration that the other Lorentz assumptions are compatible with the variation of mass with speed. Yet, at the speeds we normally encounter, the contraction of moving rods is so small that it appears extremely difficult to observe it.

A more recent experiment by Sherwyn[47] has also yielded a negative result: the author considered an elastic rod rotating about one of its ends in the laboratory frame. At low rotation rates, the length of the rod adiabatically follows the value demanded by the equilibrium lengths of the molecular bonds which, obviously, cannot be estimated by laboratory meter sticks, since they show the same dependence of length on angle. However, according to the author, at high rotation rates, when the time required to rotate 90° becomes comparable to the period of vibration of the structure, the macroscopic length should not be able to exactly follow the "bond equilibrium length."

This statement appears doubtful: if the time required to rotate 90° is comparable to the period of vibration, the adiabatic process should still apply. In all probability, only for very high rotation rates would the length of the rod not have enough time to exactly follow the "bond equilibrium length."

Note that another argument seemed, at first sight, to go against Lorentz-Fitzgerald contraction: the compressibility of matter is limited, and length contraction seems difficult to justify at very high speeds. For example at 0.9999C the ratio L/L_0 would be reduced to 1.4%. In response to this, we might say that the law was proposed following an experiment performed at low speed (the Michelson experiment). It would not take exactly the same form at very high speeds.

Today, we have come to realize that strong arguments exist in support of Lorentz-Fitzgerald contraction. One of these arguments is that Lorentz-Fitzgerald contraction allows to explain (in all directions of space, and not just in two perpendicular directions) the isotropy of the *apparent* (measured) average two-way speed of light.

We know that all the usual measurements of the *apparent* one-way speed of light by means of the Einstein-Poincaré synchronization procedure are in fact equivalent to measurements of this average velocity.[3,22] This is also true when we use clocks synchronized by slow clock transport, which is approximately equivalent to the Einstein-Poincaré procedure. (See Chapter 9.) According to Anderson, Vetharaniam and Stedman,[48] all the recent experiments purporting to illustrate the isotropy of the one-way speed of light were based on erroneous ideas because they assumed that slow clock transport allows exact synchronization. On the contrary, a number of arguments speak in favour of the anisotropy of the one-way speed of light. Although a direct measurement comes up against major difficulties, anisotropy can be deduced from the measurement of the terrestrial aether velocity, based on the fact that light signals propagate isotropically in the aether frame.

A first estimation of the absolute velocity of the solar system was made in 1968 by de Vaucouleurs and Peters, who measured the anisotropy of the extragalactic redshift relative to many distant galaxies. The experiment was repeated by Rubin in 1976. A more reliable estimate of the solar system absolute velocity was obtained by measuring the anisotropy of the 2.7° K microwave background that is uniformly distributed throughout the Universe.

> An observer moving with velocity v relative to this microwave background can detect a larger microwave flux in the forward direction ($+v$) and a smaller microwave flux in the rearward direction ($-v$). He can observe a violet shift in the forward direction ($+v$) and a red shift in the rearward direction ($-v$) (Wilhelm.)

The method has been used successively by several authors [Conklin (1969), Henry (1971), Smoot *et al.* (1977), Gorenstein and Smoot (1981), Partridge (1988).]

Another method of measurement based on the determination of the muon flux anisotropy was developed by Monstein and Wesley (1996).

A review of all these experiments has been given by Wesley[30] and Wilhelm.[30]

Marinov[15] also devoted himself to measuring the absolute velocity of the solar system by means of different devices (coupled mirrors experiment, toothed wheels experiment). The experiments are described in detail in the book by Wesley,[31] and are quoted by Wilhelm.[30]

Many physicists were suspicious of Marinov's claims. They regretted not having the proof that the experiments were actually performed. But Wesley, who knew him very well, attested that he had no reason to doubt his reported results. According to Wesley, "Marinov was extremely reliable and scrupulously honest in all of his personal dealings with people. He had a firm grasp on reality."[53] The author of the present text can personally attest that he possesses a photograph of the toothed wheels device.

According to Wesley, "the Marinov (1974, 1977a, 1980b) coupled mirrors experiment, is one of the most brilliant and ingenious experiments of all time. It measures the very small quantity v/C, where v is the absolute velocity of the observer, by using very clever stratagems."

The coupled mirrors experiment demonstrated that the absolute velocity of the solar system v, is of the order of 300 ± 20 km/sec, and that the speed of light is $C - v$ in the direction of motion of the solar system, and $C + v$ in the opposite direction. (Note that the orbital motion of the Earth around the Sun is far slower (about 30 km/sec), and that the rotational motion at the latitude of the experiment was of the order of 0.5 km/sec.)

All the above mentioned experiments proved in agreement with the results obtained by Marinov. They demonstrated the existence of a fundamen-

Figure 11. The path of the light signal along the arm perpendicular to the direction of motion viewed by an observer from the aether frame.

tal frame, whose absolute speed is zero, but whose relative speed with respect to the Earth frame is of the order of 300 km/sec.

Present-day arguments for Lorentz contraction

Michelson experiment

Consider now a Michelson interferometer whose longitudinal arm is aligned along the x_0-axis of a system of co-ordinates S_0 $(0, x_0, y_0, z_0)$ attached to the Cosmic Substratum. The arm is at rest in the Earth frame, which moves along the x_0-axis at speed v. It is easy to verify that, in reply to the statement that the speed of light is $C-v$ in the $+x_0$ direction, and $C+v$ in the opposite direction, the arm will be contracted in the ratio

$$\ell = \ell_0 \sqrt{1 - v^2/C^2}, \qquad (6.1)$$

where ℓ is the length of the arm in motion, and ℓ_0 the length at rest.

Using the same experimental device we can demonstrate that the *apparent* (measured) average two-way speed of light along the x_0-axis, is equal to C, independently of the speed v. Let us demonstrate formula (6.1).

We do not know *a priori* if $\ell = \ell_0$ or not. The two-way transit time of light along the longitudinal arm will be:

$$t_1 = \frac{\ell}{C-v} + \frac{\ell}{C+v} = \frac{2\ell}{C(1-v^2/C^2)}. \qquad (6.2)$$

Now, in the arm perpendicular to the direction of motion, there is no length contraction. The speed of light is C exclusively in the aether frame. The signal propagates from a point P in this frame toward a point $\underline{O}$ at the extremity of the arm, and then comes back toward point P'. During that time, the interferometer has covered the path vt_2. (See Figure 11.)

We have:

$$\left(C\frac{t_2}{2}\right)^2 - \left(v\frac{t_2}{2}\right)^2 = \ell_0^2,$$

$$\ell_0 = \frac{t_2}{2}\sqrt{C^2 - v^2},$$

so that

$$t_2 = \frac{2\ell_0}{C\sqrt{1 - \frac{v^2}{C^2}}}. \tag{6.3}$$

Neglecting the tiny displacement of the fringes observed when we change the orientation of the apparatus, which is really too small to explain the existence of an aether wind of the order of 300 km/sec, we can write $t_1 = t_2$, that is:

$$\frac{2\ell}{C(1 - v^2/C^2)} = \frac{2\ell_0}{C\sqrt{1 - v^2/C^2}}.$$

Hence

$$\ell = \ell_0\sqrt{1 - v^2/C^2}.$$

Therefore, if we take the anisotropy of the one-way speed of light into account, length contraction must no longer be considered an *ad hoc* hypothesis. On the contrary, it must be seen as a necessary cause of the Michelson result.

Now, on account of clock retardation, the *'apparent'* two-way transit time of light will be (from 6.3):

$$\frac{2\ell_0}{C}.$$

Since the length of the longitudinal arm is determined with a contracted standard, it is found equal to ℓ_0 and not to ℓ, so that the *'apparent'* average two-way speed of light along the x_0-axis will be found equal to C. It is in fact different from its real value, which according to formula (6.2) is $C(1 - v^2/C^2)$.

N.B. In the absence of length contraction, the average two-way speed of light would not have been found equal to C, in contradiction with the experiment.

Average two-way speed of light

But this is not all. We will now demonstrate that length contraction implies the independence of the *'apparent'* average two-way speed of light from any direction in space and from the speed v. The demonstration is based on Builder and Prokhovnik's[22] studies, whose importance is indisputable.

However, as we will see, some of the conclusions drawn by Prokhovnik were questionable and could not enable us to demonstrate that this '*apparent*' velocity is equal to C in any direction of space.

Consider the two inertial frames discussed in the previous paragraph, and suppose that a rod AB making an angle θ with the x_0, x-axis, is at rest with respect to frame S. (See Figure 12.) At the two ends of the rod, we place two mirrors facing one another, perpendicular to the axis of the rod $\ell = AB$. At the initial instant, the two frames S_0 and S are coincident. At this very instant a light signal is sent from the common origin and travels along the rod toward point B. After reflection the signal returns to point A.

We do not assume *a priori* that $\ell = \ell_0$ (where ℓ_0 is the length of the rod when it is at rest in frame S_0).

We first note that the path of the light signal along the rod is related to the speed C_1 by the relation.

$$C_1 = \frac{AB}{t},$$

where t is the time needed by the signal to cover the distance AB. (See Figure 13.)

In addition, when the signal reaches point B, frame S has moved away from frame S_0 a distance: $AA' = vt$, so that

$$v = \frac{AA'}{t}.$$

Now, from the point of view of an observer at rest in frame S_0, the signal goes from point A to point B' (Figure 13). C being the speed of light in frame S_0, we therefore have

$$\frac{AB'}{t} = C,$$

and hence, the projection of the speed of light in frame S along the x-axis will be:

$$C \cos \alpha - v.$$

We remark that

$$C \cos \alpha - v = C_1 \cos \theta.$$

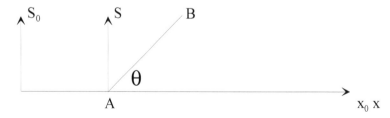

Figure 12 - The rod AB is at rest with respect to frame S.

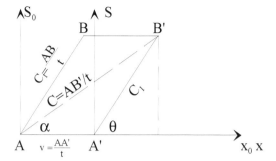

Figure 13. The speed of light is equal to C_1 from A' to B', and to C from A to B'.

Since the three speeds, C, C_1 and v are proportional to the three lengths AB', AB and AA' with the same coefficient of proportionality, we have

$$C^2 = (C_1 \cos\theta + v)^2 + C_1^2 \sin^2\theta,$$

and thus

$$C_1^2 + 2vC_1 \cos\theta - (C^2 - v^2) = 0. \quad (6.4)$$

(It should be emphasized that, in equation (6.4), the three speeds C, C_1 and v have been measured with clocks attached to frame S_0, see below.)

Solving the second degree equation, we obtain:

$$C_1 = -v\cos\theta \pm \sqrt{C^2 - v^2 \sin^2\theta}.$$

The condition $C_1 = C$ when $v = 0$ compels us to only retain the "+" sign so:

$$C_1 = -v\cos\theta + \sqrt{C^2 - v^2 \sin^2\theta}. \quad (6.5)$$

The return of the light signal is illustrated by Figure 14. From the point of view of the observer in frame S, the light comes back to its initial position with the speed C_2. Therefore we can write:

$$C_2 = \frac{B'A'}{t'}.$$

For the observer in frame S_0, the light travels from B' to A'' with speed C,

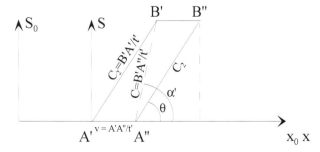

Figure 14 - The speed of light is equal to C_2 from B'' to A'', and to C from B' to A''.

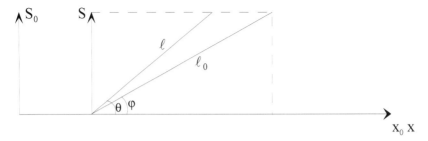

Figure 15. Along the x_0x-axis, the projection of the rod ℓ_0 contracts. Along the y-axis it is not modified.

so that
$$C = \frac{B'A''}{t'}.$$

During the light transfer, frame S has moved from A' to A'' with the speed v, so that:
$$v = \frac{A'A''}{t'}.$$

The projection of the speed of light along the x-axis will be
$$C_2 \cos\theta = C \cos\alpha' + v.$$

We easily verify that:
$$(C_2\cos\theta - v)^2 + (C_2\sin\theta)^2 = C^2,$$

and finally
$$C_2 = v\cos\theta + \sqrt{C^2 - v^2 \sin^2\theta}. \qquad (6.6)$$

The two-way transit time of light along the rod (measured with clocks attached to frame S_0) is thus:
$$2T = \frac{\ell}{C_1} + \frac{\ell}{C_2}. \qquad (6.7)$$

According to the experiment, T must be essentially independent of angle θ. Therefore $2T$ must be equal to
$$\frac{2\ell_0}{C\sqrt{1 - v^2/C^2}},$$

which is the two-way transit time along the y direction (calculated above.)

We can see that, in order for this condition to be satisfied, the projection of the rod along the x-axis must shrink in such a way that:
$$\ell \cos\theta = \ell_0 \cos\varphi \sqrt{1 - v^2/C^2}. \text{ (See Figure 15.)}$$

where φ was the angle separating the rod and the x_0-axis when the rod was at rest in frame S_0. From:

$$\ell_0 \cos \varphi = \frac{\ell \cos \theta}{\sqrt{1 - v^2/C^2}},$$

and

$$\ell_0 \sin \varphi = \ell \sin \theta,$$

we easily verify that:

$$\left(\frac{\ell \cos \theta}{\sqrt{1 - v^2/C^2}} \right)^2 + (\ell \sin \theta)^2 = \ell_0^2.$$

And finally:

$$\ell = \frac{\ell_0 \left(1 - v^2/C^2\right)^{1/2}}{\left(1 - v^2 \sin^2 \theta / C^2\right)^{1/2}}. \tag{6.8}$$

Replacing ℓ with this expression in (6.7) we obtain, as expected:

$$2T = \frac{2\ell_0}{C\sqrt{1 - v^2/C^2}},$$

and the average transit time of light along rod ℓ will be

$$T = \frac{\ell_0}{C\sqrt{1 - v^2/C^2}}. \tag{6.9}$$

Finally, from the above experiments demonstrating the anisotropy of the one-way speed of light, we have deduced that the projections of this speed along the x-axis of frame S are $C \cos \alpha - v$ when the light runs in the forward direction and $C \cos \alpha' + v$ when it runs in the opposite direction. These results, combined with the isotropy of the two-way transit time of light, are sufficient to show that the projection of the rod along the x-axis shrinks in such a way that

$$\ell \cos \theta = \ell_0 \cos \varphi \sqrt{1 - v^2/C^2}.$$

But that is not all. The same conditions, combined with clock retardation, allow us to demonstrate that the *'apparent'* (experimental) average two-way speed of light is isotropic.

Clock retardation is an experimental fact. Let us designate the *apparent* average transit time in frame S along the rod as ε. We will have (from (6.9)):

$$\varepsilon = T\sqrt{1 - \frac{v^2}{C^2}}$$
$$= \frac{\ell_0}{C}.$$
(6.10)

Now, the length of the rod, measured with the contracted meter stick in frame S, is always found equal to ℓ_0, so that the average two-way speed of light is, erroneously, found equal to C in any direction of space and independently of speed v. This result is highly meaningful, and is a direct consequence of the facts deduced from the Michelson and Morley and Marinov experiments (or any other equivalent experiment).

It is worth noting that in our demonstration, although we are indebted to Prokhovnik, we differ with his conclusions.[22] Indeed, since $C = AB'/t$ and also $C = B'A''/t'$, it is obvious that t and t' are the real transit times of light along the rod (as measured with clocks attached to the aether frame.) Now, since $C_1 = AB/t$ and $C_2 = B'A'/t'$, there is no doubt that C_1 and C_2 are also measured with clocks attached to frame S_0. This is also the case for $2T = (\ell/C_1) + (\ell/C_2)$.

Nevertheless, in his book *The logic of special relativity*,[22] in the Chapter "The logic of absolute motion," Prokhovnik identifies the time $2T = 2\ell_0/C\sqrt{1 - v^2/C^2}$ with the two-way transit time of light along the rod, as measured with clocks attached to the moving frame. (See Prokhovnik's formula 5.2.4.) This cannot be true for the reason indicated above.

(Note that in our notation the moving frame is designated as S, while in Prokhovnik's notation, S designates the aether frame and A the moving frame. We will continue the demonstration with our own notation.)

In addition, if Prokhovnik's approach were true, the *'apparent'* two-way speed of light measured with a standard in frame S would not be C. Indeed, since this standard is also contracted, observer S would find ℓ_0 for the length of the rod.

Therefore, the *'apparent'* average two-way speed of light in frame S would have been:

$$\frac{2\ell_0}{2\ell_0/C\sqrt{1-v^2/C^2}} = C\sqrt{1-v^2/C^2},$$

which is not in agreement with the experimental facts.

The real two-way transit time of light along the rod, as measured with clocks attached to frame S_0, is in fact $2\ell_0/C\sqrt{1-v^2/C^2}$, and the two way clock display (measured with clocks attached to frame S), is $2\ell_0/C$. This corresponds to the experimental facts, since, with these values, the 'apparent' average two-way speed of light in frame S is found equal to

$$2\ell_0 \bigg/ \frac{2\ell_0}{C} = C.$$

(For more detailed explanations see also Chapter 3 of the present text.) Note also that the real, average two-way speed of light along the x_0, x-axis is

$$\frac{2\ell_0\sqrt{1-v^2/C^2}}{2\ell_0/C\sqrt{1-v^2/C^2}} = C(1-v^2/C^2),$$

which, as expected, tends to 0 when $v \Rightarrow C$.)

Remark

We could have obtained the same result without length contraction by assuming that the speed of light in direction AB was (from formula (6.8)

$$C'_1 = C_1 \times \frac{\left(1-v^2\sin^2\theta/C^2\right)^{1/2}}{\left(1-v^2/C^2\right)^{1/2}}, \qquad (6.11)$$

and in the opposite direction

$$C'_2 = C_2 \times \frac{\left(1-v^2\sin^2\theta/C^2\right)^{1/2}}{\left(1-v^2/C^2\right)^{1/2}}, \qquad (6.12)$$

giving

$$\frac{\ell_0}{C'_1} + \frac{\ell_0}{C'_2} = 2T = \frac{\ell}{C_1} + \frac{\ell}{C_2}.$$

Let us try to verify this hypothesis.

Consider the case where $\theta = 0$ in formulas (6.11) and (6.12). The speed of light would then have been $(C-v)/(1-v^2/C^2)^{1/2}$ in the direction AB and $(C+v)/(1-v^2/C^2)^{1/2}$ in the opposite direction. It is obvious that the factor $1/(1-v^2/C^2)^{1/2}$ would have resulted from the proximity of the celestial bodies, since, at some distance from them, in a reference frame not influenced by the presence of matter, there is no possible explanation for the existence of this term.

Because this factor is multiplicative, it indicates an increase of the speed of light in the neighbourhood of celestial bodies, which in turn implies a partial dragging of the aether. This hypothesis cannot be justified for the following reasons:

1. The experiment by Lodge[52] demonstrated that the speed of light is not changed in the neighbourhood of a rotating wheel.
2. A partial dragging of the aether would have increased the speed of light in the direction of motion of the Earth and reduced it in the reverse direction. And the multiplicative factor would have been different in the two opposite directions, which is not the case here.
3. On the contrary, there are no logical obstacles for Lorentz contraction.

Moreover, there are theoretical arguments in favour of the process:

a. Since the anisotropy of the one-way speed of light is proved, we are limited to two hypotheses (as we saw earlier in the text) and we must choose between them. Elimination of the above-mentioned hypothesis renders length contraction necessary.
b. According to Wilhelm: "Since matter consists of positive (nuclei) and negative (electrons) charges, the contraction of their equipotential surfaces causes Lorentz-Fitzgerald's contraction."
c. The above-mentioned experiments, which demonstrate the anisotropy of the one-way speed of light, provide a physical basis for the theoretical explanations of Heaviside, Larmor, Lorentz, Bell, Prokhovnik and several prominent modern physicists, who assume length contraction. It invalidates the other theories, such as the Ritz theory (ballistic), Einstein theory, Stokes theory (completely dragged aether), and Abraham theory (seen above.)

Note

The hypothesis of the aether dragged by the Earth has generally been rejected because it seemed in contradiction with the theory of aberration. However, as demonstrated by Beckmann,[49] Mitsopoulos[50] and Makarov,[51] this is not the case.

Nevertheless, the dragged aether theory is contradicted by the experiment of Lodge,[52] who demonstrated that the speed of light is not modified in the neighbourhood of a rotating wheel; and by the experiments of Marinov[15,31] (toothed wheels experiment and coupled mirrors experiment,) or any other experiment demonstrating the anisotropy of the one-way speed of light.

Chapter 7

Implications for Fundamental Physics

In light of the new data presented in this manuscript, it is appropriate to review certain principles of physics considered as fundamental.

The relativity principle

It is generally felt that the conceptual content of this principle is well defined. Nevertheless, this is not entirely exact; in fact, we can distinguish at least three definitions of the relativity principle, whose authors were Galileo, Poincaré and Einstein, respectively.

Galileo's relativity principle

Although not defined as such, the idea of relativity occupied such great philosophers as Jean Buridan, rector of the University of Paris (1300-1358), Giordano Bruno (1568-1600), Descartes (1596-1650), Leibniz (1646-1716), and Newton (1642-1727). But Galileo was the first to give the idea a clear formulation.

Aristotelian physics treated motion and rest as absolute and completely distinct, with the Earth, in a state of absolute rest, occupying the centre of the Universe. Galileo departed completely from this attitude. For him, uniform motion presents no absolute character. Any object at rest with respect to a given reference system is, at the same time, in motion with respect to another reference system. Rest and motion are not fundamentally different; they have only relative character.

As an example of his relativity principle, Galileo cited the case of a stone released from the top of a ship's mast as the ship sails uniformly in a straight line. If, according to Galileo, motion had an absolute character, the stone would fall at a distance from the foot of the mast. Yet the stone drops precisely at the foot of the mast, a result he interpreted as due to the fact that the boat is at rest in its inertial system.

This argument is actually questionable. Indeed, the stone has momentum mv parallel to the direction of motion of the boat. The fact that it is released does not eliminate this momentum, which constrains it to continue its horizontal motion along with the boat, while its vertical motion is determined by the law of gravitation.

Therefore, even if motion and rest do present an absolute character, the stone will still drop at the foot of the mast. (Note that our reasoning is entirely exact only in the ideal case where the motion of the boat is rectilinear and uniform and where air resistance is negligible. To be rigorous we should also make allowance for the resistance opposed by the substratum. But in the present example this would be very small.)

The fact remains that it is to Galileo's credit that the meaning of motion is clearly understood. Contrary to the Aristotelian belief, Galileo demonstrated that, while motion needs a motor to be produced, it does not need a motor to be maintained.

Poincaré's relativity principle

The Galilean relativity principle was limited to the laws of mechanics and to bodies moving at low speeds. Poincaré's goal was to extend it to all the laws of physics, and in particular to Maxwell's electromagnetism, which seemed to be an exception to the rule.

In order to bring the laws of electromagnetism back into line, Poincaré had to resort to a set of equations he baptized the "Lorentz transformations," which should constitute a group. Poincaré expressed his principle in the following terms:

> It seems that the impossibility of observing the absolute motion of the Earth is a general law of nature. We are naturally inclined to admit this law that we shall call the relativity postulate, and to admit it without restriction.[18]

This sentence does not imply dismissal of absolute motion, but rather the impossibility of observing it. However, on another occasion, Poincaré stated:

> …there is no absolute space. All the motions that we can observe are relative motions.[18]

Nevertheless, Poincaré did not reject the aether. His acceptance of this medium is expressed in several quotations. (See for example his Lille lecture in 1909.[18])

Poincaré's point of view cannot be accepted. We have demonstrated this in the example of two vehicles moving in opposite directions toward two symmetrical targets. (See end of Chapter 2 of this text.) In addition, since Poincaré assumes Lorentz contraction, a rod in motion with respect to the aether frame will shrink when its orientation is changed (from perpendicular to parallel). Of course, we will not be able to measure this contraction with a meter stick since the stick also shrinks in the same proportion. However, at high speed, the process will be observable and it can inform us about the absolute motion of the rod. (Note too that, with four meter sticks in motion at high speed, two of them being parallel to the direction of motion and other two perpendicular, we will construct a rectangle instead of a square.)

Therefore, contrary to Poincaré's view, the Lorentz aether is not compatible with the relativity principle.

Einstein's relativity principle

In its original formulation, Einstein's relativity principle rejects the aether hypothesis. It can be expressed as follows: "All inertial frames are equivalent. The laws of nature, including those of electromagnetism, take the same form in all of these frames."

As we saw in Chapter 2 and in the conclusion of Chapter 4, Einstein's relativity principle, like Poincaré's, leads to results conflicting with logic.

Concluding remarks

Generally speaking, Galileo's relativity idea (as well as those of Poincaré and Einstein which derive from it) cannot be retained because it implies reciprocity, and this is in contradiction with mass-energy conservation. If the relativity principle were true, the substratum should exert an identical influence on all inertial systems[*] or no influence at all, because if this were not the case, the physical laws would appear different in two inertial frames, S_0 and S. (This notion appears implicit in modern aether models.)

Now, suppose that a space-ship leaves S_0 and, after acceleration, reaches a constant speed v and becomes firmly attached to frame S. We will suppose that it has used energy E_C. If all inertial frames were equivalent, the frames S_0 and S would only be distinguished from one another by their relative speed. Therefore, in order to come back to S_0 the space-ship should use the same energy as it does going from S_0 to S. And this would be true no matter what point in S_0 the space-ship reaches upon its return. (Indeed, a body at rest with respect to a given inertial frame has a well defined mass-energy, whatever its position may be in this inertial frame. This mass-energy is equal to the sum of the internal mass-energy of the body $m_0 C^2$ and its kinetic energy, which is the same in any position in the inertial frame.[**]) Therefore, the spaceship would have used energy $2E_C$ to leave and then recover the same energy state, and there would be no conservation of mass-energy.

---

[*] This is what Einstein means when in the conclusion of his book *Ether and Relativity* he claims... this ether may not be thought as endowed with the quality characteristic of ponderable media, as consisting of parts which may be tracked through time. The idea of motion may not be applied to it. A Einstein, *Sidelights on relativity*, Dover, New York.

[**] Of course this last statement applies exactly only in an ideal inertial frame where there is no gravity and where the motion of the frame is strictly rectilinear and uniform.

Consequently, on account of the assumed complete reciprocity of frames S_0 and S, the relativity principle implies that the space-ship must accelerate in the same way to pass from S to S_0 as it does from S_0 to S.

Yet, in order to come back to S_0, the spaceship has only to restore energy E_C to the environment, and the law of conservation of mass-energy will be obeyed. The situation is similar to that of a body which acquires potential energy E when it moves from one level A to another level B. Upon its return to A, the body must give up the same energy E.

The point of view of the fundamental aether theory is therefore completely different. Here there is no identity between S_0 and S. From S_0 to S the space-ship must accelerate, and hence we must supply energy to it; from S to S_0 it must decelerate and give up the same energy to the environment. Therefore, the frames S_0 and S are not equivalent. But this implies that S_0 and S are affected differently by the substratum.

Since we must use energy to go from S_0 to S, we can conclude that the substratum offers resistance to motion which increases with absolute velocity, and which will be higher in S than in S_0.[*] A consequence of this is that the principle of inertia cannot be exactly obeyed. (See below.)

Furthermore, the relativity principle implies that the available mass-energy of a body A is not well defined. It has only relative value with respect to another body B, and if the speed of body B tends toward the speed of light, the kinetic energy of A with respect to B will tend toward infinity. This is untenable. The total available mass-energy of body A is finite, and is defined with respect to the fundamental frame. It is absurd to consider that it depends on the speed of another body.

Therefore, the relativity principle cannot be maintained. It nevertheless remains that the uniform motion of a given inertial frame is imperceptible to an observer at rest in this frame. But this does not result from the fact that only relative speeds have a meaning, or from the Galilean idea that "motion is like nothing" (*i.e.*, essentially relative). Rather, it is due to the fact that only variations of absolute speeds are perceived. Uniform absolute speed is not perceived precisely because it remains unchanged all the time, *i.e.*, because the energy of the body in motion is not modified.

Important note

As we have seen, the relativity principle implies that two identical bodies attached to any two inertial systems S_1 and S_2, assume a completely symmetrical situation, and therefore possess the same energy status.

If a spaceship needs to use energy to move from S_1 to S_2, it will also use the same energy to move from S_2 to S_1 and the energy will not be con-

---

[*] An important consequence of this is that, except for the fundamental frame, real frames are never exactly inertial. At low speeds they can be considered as almost exactly inertial. But as absolute velocity tends towards the speed of light, the inertial character is lost.

served. In order for the energy to be conserved, it would have to move from S_1 to S_2 (or from S_2 to S_1) without consuming energy.

Therefore, assuming that the rest mass is m_0 we should have:
$$(m - m_0)C^2 = 0$$
$$\Rightarrow m_0 C^2 \left[\left(1 - v^2/C^2\right)^{-1/2} - 1 \right] = 0,$$

where m is the mass of the spaceship in S_2 viewed from S_1, and v the speed of S_2 with respect to S_1. Since $v \neq 0$ the equation implies that $m_0 = 0$. So, paradoxically, the fact that the aether exerts an identical influence (or no influence at all) on all inertial frames would imply that the bodies do not possess mass.

On the other hand, the fundamental aether theory implies a different influence of the substratum on S_1 and S_2. If we suppose that v_1 is the speed of S_1 with respect to the aether frame S_0 and v_2 is the speed of S_2 with respect to S_0, the energy acquired by the body when it moves from S_1 to S_2 will be

$$m_0 C^2 \left[\left(1 - v_2^2/C^2\right)^{-1/2} - \left(1 - v_1^2/C^2\right)^{-1/2} \right] \neq 0,$$

and as a consequence $m_0 \neq 0$.

We can conclude that the existence of a mass $m_0 \neq 0$ depends not only on the quantity of matter, but also on the action of the aether on physical bodies, which implies that the relativity principle is incompatible with the existence of mass.

The Higgs field hypothesis should be addressed from this new standpoint.

Mass-energy conservation

This law must be viewed as unquestionable because mass and energy cannot arise from nothing and, conversely, they cannot be destroyed. Nevertheless, in processes where mass or energy are exchanged, we must allow for the fact that the aether can absorb or supply part of the energy.

Principle of inertia

In a previous paper,[5] we stated that any objection to the relativity principle, implies a challenge to the principle of inertia. It is important here to provide further information in order to specify what we mean. In its original formulation, the principle was expressed in concrete terms: "a marble sliding on a perfectly smooth horizontal surface (without any friction) *in vacuo* remains perpetually in its state of motion."

Of course, if, in agreement with the Galilean relativity principle, rest and uniform motion are only relative, we can view the marble as at rest in its reference system. As a result, it must remain in this state of rest. But in the fundamental aether theory proposed here, absolute rest exists and is distinct from motion. The difference results from the existence of the aether. Under the action of the aether, the marble will experience a gradual slowing down, hardly perceptible, but not null. The Galilean principle of inertia is therefore challenged.

Now, in its modern sense, the principle of inertia can be expressed as: "a body not subjected to any external force remains perpetually in its state of motion." If we assume this more precise definition, the principle of inertia appears as essentially correct. The aether should exert a pressure—very weak, but which increases with velocity—on the body, causing the body to slow down. If this pressure were balanced, the body would remain in its state of motion. This condition appears necessary in order for the law of mass-energy conservation to be obeyed.

Conservation of momentum

Insofar as the particles interacting in a collision are slowed down by the aether drift, their total quantity of motion cannot be the same before and after the collision. This effect, imperceptible at low speed, cannot be ignored at high speed ($v > 10^5$ km/sec).

In all probability, if the pressure exerted by the aether were balanced, the total quantity of motion would be conserved.

Mass-energy equivalence

In the relativity theory, whatever inertial frame S is considered, the energy content of a body of mass m_0 measured by an observer at rest in this frame, is $E_0 = m_0 C^2$, where m_0, the rest mass, is the same in all inertial frames. This implies that when a body moves from one inertial frame S_1 to another frame S_2 receding from S_1 at speed v, its rest energy remains unchanged. The fact that the absolute kinetic energy of the body has increased has, nevertheless, not changed its rest mass and its rest energy.

This viewpoint is simply untenable. Of course we are aware that, if we use a standard that also passes from S_1 to S_2 to measure the mass of the body by comparison, the rest mass will be (erroneously) found identical in S_1 and S_2. This is because the rest mass of the standard will have changed in the same ratio as the body's mass. But the real rest masses in S_1 and in S_2 are different. (See Chapter 8.)

Thus, even though they are difficult to estimate, the rest mass and rest energy of a body increase when the speed of the body increases with respect to the fundamental frame.

Variation of mass with speed

Contrary to special relativity, the fundamental aether theory developed here attributes a different mathematical form to the law of variation of mass with speed, according to whether the mass of the moving body is compared to its mass in the fundamental frame or to that in any other inertial frame. (See Chapter 8.) Accordingly, this result is once more at variance with the relativity principle.

Invariance of the one-way speed of light

This so-called postulate comes up against a number of difficulties that have been analysed in the body of the text. (See Chapter 2.) Contrary to what is often believed, what we actually measure is not the one-way speed of light, but the *apparent* average two-way speed of light altered by systematic errors of measurement. It is this quantity which is invariant.

Twin paradox

Let us imagine two spaceships piloted by two twins which recede uniformly and symmetrically from a point O of the Cosmic Substratum. After covering the distance D (and $-D$) along the same straight line, their clock displays will be identical.

Conversely, if one space-ship remains in the preferred inertial frame, and the other covers the distance $2D$, their clock displays will actually be different. There is no paradox, since one of the twins is in a state of absolute rest, while the other is in a state of absolute motion, and there is no reciprocity.

Nothing similar can exist in Einstein's special relativity. Relativity theory does not assume the existence of a privileged frame.

This also demonstrates that, contrary to Poincaré's opinion, the existence of an aether frame is not compatible with the relativity principle, and that the fundamental aether theory leads to results completely different from special relativity.

Relativity of time

In the fundamental aether theory, clocks in motion with respect to the fundamental frame slow down; but conversely, clocks at rest in the fundamental frame are not affected by the relative motion of bodies. Rather than a relativity of time, the effect is clock retardation. There is no real reciprocity. (Nevertheless, as demonstrated by Prokhovnik, on account of the Einstein-Poincaré synchronization procedure, *apparent* reciprocity can take place in some particular situations.)[3,22]

This completely distinguishes the theory presented here from relativistic theories, which admit perfect reciprocity of observations.

Relativity of simultaneity

Relativity of simultaneity must be distinguished from relativity of time. As seen in Chapter 2, relativity of simultaneity appears completely conventional and depends on the method of clock synchronization. Perfect simultaneity can be defined[4,6] by means of a more appropriate method of synchronization.

Minkowski space-time

Space-time transformations are nothing more than the Galilean transformations disguised by systematic errors of measurement. After correction for these errors, the true transformations in which space and time are distinct are recovered. Minkowski's space-time is therefore reduced to an artefact resulting from a convention of synchronization based on questionable assumptions.

Length contraction

According to special relativity, length contraction is observational and reciprocal. That is to say, seen from a reference frame A, a rod at rest in a reference frame B will appear contracted. But seen from B, the same rod at rest in frame A will appear contracted in the same proportion.

When it is attached to a given inertial frame, for an observer at rest in the frame, the rod will have the same length no matter which inertial frame we consider. This length is defined as the rest length of the body. Now, insofar as the relativity principle has been refuted, this viewpoint cannot be maintained.

According to the fundamental aether theory proposed in this text, the rod is really contracted. The amount of this contraction depends on the speed of the rod with respect to the aether frame. This contraction will be the same for any observer, no matter what the speed of the observer with respect to the rod. This will also be true for an observer at rest relative to the rod.

Of course, with these assumptions, the contraction cannot be measured with a ruler that has passed from the aether frame S_0 to the moving frame S_1 at the same time as the rod, since the ruler will be contracted in the same ratio as the rod. As a result, a number of physicists believe that real and non-reciprocal length contraction is compatible with the relativity principle. If the length appears the same in all inertial frames, they think that it cannot tell us anything about the absolute motion of a given inertial frame with respect to the substratum. This was Poincaré's point of view, for example.

But the argument can be refuted, as follows. In a frame moving at high speed, if we try to construct a square with four meter sticks that were identical in the aether frame (two of them being parallel to the direction of motion and the other two perpendicular), the figure will be seen as a rectangle, the surface of which will depend on the speed with respect to the aether frame. Contrary to Poincaré's opinion, real and non-reciprocal length contraction is not compatible with the relativity principle.

But, as we saw in Chapter 6, a number of arguments today support Lorentz contraction.

Chapter 8

Mass and Energy in the Fundamental Aether Theory

The fundamental extended space-time transformations derived in the previous chapters conceal hidden variables. After correction of the systematic errors of measurement due to length contraction, clock retardation and imperfect clock synchronization, they reduce to the Galilean relationships, $x' = x \pm vt$, $t' = t$.

As we saw at the beginning of Chapter 4, this result implies that the total relativistic quantity of motion of particles interacting in a collision is not exactly conserved in all inertial frames. Therefore, the so called "law of conservation of the total relativistic momentum" cannot be used to demonstrate or to disprove $m = m_0 \gamma$. We shall nevertheless demonstrate, by means of arguments independent of relativity, that the law $m = m_0 \gamma$ applies. But, contrary to relativity, m_0 is the rest mass in the fundamental frame; it is not the rest mass in all inertial frames.

For the following it is important to note that, in the particular cases where the total quantity of motion is conserved, the conservation must be effective for all observers, no matter whether they are at rest or in motion with respect to the system in which the collision occurs. (See the example given below.) This fact will be used to demonstrate the law $E = mC^2$ without relying on relativistic arguments. We will then deduce $m = m_0 \gamma$.

Demonstration of $E = mC^2$ without relativistic arguments

Consider a body at rest in the fundamental frame S_0 that emits N identical

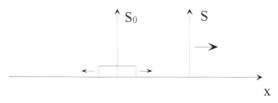

Figure 16. The body at rest in frame S_0 emits N identical photons in two opposite directions.

photons simultaneously in two opposite directions (+x and –x,). See Figure 16. (For this demonstration, we will follow arguments given by Rohrlich,[54] but with different assumptions.)

Consider now another inertial frame S moving along the x-axis at speed v, (with $(v/C)^2 \ll 1$.) In frame S_0, the total momentum is conserved. This must also be true for any observer moving with respect to frame S_0. With respect to frame S, we have:

$$P_0 = P_1 + N\frac{hv}{C}\left(1+\frac{v}{C}\right) - N\frac{hv}{C}\left(1-\frac{v}{C}\right),$$

where P_0 is the initial momentum, and P_1 the final momentum of the body. The other terms are the momenta of the photons altered by the Doppler shift. (Note that $p = hv/C$ is a formula from classical electrodynamics, independent of relativity.)

Viewed from frame S, the momentum $\Delta(mv)$ lost by the body will be:

$$P_0 - P_1 = 2N\frac{hv}{C^2}v.$$

Since, obviously, the source is at rest in frame S_0 both before and after emission, with respect to frame S it must have the speed v both before and after emission, so:

$$\Delta(mv) = v\Delta m = \frac{2Nhv}{C^2}v. \qquad (8.1)$$

Now, according to the energy conservation law:

$$E_0 = E_1 + Nhv\left(1+\frac{v}{c}\right) + Nhv\left(1-\frac{v}{c}\right) = E_1 + \Delta E$$
$$= E_1 + 2Nhv. \qquad (8.2)$$

$\Delta E = 2Nhv$ is the variation of energy resulting from the emission of the photons. From (8.1) and (8.2) we obtain

$$\Delta E = \Delta mC^2.$$

Note that this mass-energy equivalence formula has been obtained without the help of the Lorentz-Poincaré transformations.

Variation of mass with speed

Consider a body *at rest in the fundamental frame*, which is subjected to a force F. The elementary expression for the kinetic energy is.

$$dE_C = Fd\ell,$$

where $Fd\ell$ is the work carried out by the force F in the displacement $d\ell$. (We suppose that F and $d\ell$ are aligned.)

This expression can also be written as follows:

$$\frac{dE_C}{dt} = Fv.$$

From $F = dp/dt$ we obtain

$$\frac{dE_C}{dt} = v\frac{dp}{dt},$$

From this expression we easily deduce:

$$\frac{dE_C}{dv}\dot{v} = \frac{dp}{dv}v\dot{v}.$$

We can finally verify that

$$\frac{dE_C}{dv} = v\frac{dp}{dv}. \tag{8.3}$$

This expression, which connects the kinetic energy and the momentum, was derived by Lewis[55,37] in 1908.

Now, as seen previously, the equivalence of mass and energy takes the form

$$E = mC^2 = E_C + m_0 C^2. \tag{8.4}$$

From the expression for momentum

$$p = mv, \tag{8.5}$$

and the expression for energy, we can write:

$$p = \frac{E}{C^2}v.$$

From (8.3) and (8.4) we have:

$$\frac{dE}{dv} = v\frac{dp}{dv}.$$

Replacing E and p by their expressions given in (8.4) and (8.5), we obtain

$$\frac{dm}{dv}C^2 = \frac{dm}{dv}v^2 + mv,$$

so

$$\frac{dm}{m} = \frac{v}{C^2 - v^2}dv.$$

Designating $C^2 - v^2$ as u so that $v\,dv = -du/2$, we then find

$$\log m = -\frac{1}{2}\log(C^2 - v^2) + \log k$$

$$= \log k(C^2 - v^2)^{-1/2}.$$

and

$$m = \frac{k}{C\sqrt{1 - v^2/C^2}}.$$

For $v = 0 \Rightarrow$, $m = k/C = m_0$, thus:

$$m = \frac{m_0}{\sqrt{1-v^2/C^2}}, \qquad (8.6)$$

where m_0 is the rest mass. As we will now see, expression (8.6) is completely exact only when m_0 is the mass at rest in the fundamental frame.

Variation of mass with speed in relativity and in the fundamental aether theory

In relativity, since no absolute frame exists, the mass of a body attached to a given inertial frame, viewed by an observer at rest in this frame, is always identical. This mass is defined as the proper mass or the rest mass of the body.

If the body moves with respect to a reference frame S with velocity v, its mass with respect to S is assumed to be:

$$m = \frac{m_0}{\sqrt{1-v^2/C^2}},$$

whatever the reference frame S may be.

The point of view of the fundamental aether theory is completely different. Consider a body having mass m_0 in the fundamental frame S_0. Since this body needs to acquire kinetic energy E_C in order to move from S_0 to any other inertial frame S, the rest mass of the body in frame S will be $m_0 + E_C/C^2$. This means that a hierarchy of rest masses exists, each a function of the absolute speed of the body.

(Note that it is necessary to distinguish the real mass from the measured mass, which may be incorrectly determined. If we measure the mass m_0 of a body in the fundamental frame S_0 by comparison with a standard μ_0 and if m_0 and μ_0 are transported into another inertial frame S, they are changed in the same ratio. As a result, the mass m_0 appears not to have changed, which is incorrect.)

In other words, the real mass m of the body in frame S, cannot be measured by an observer at rest in this frame. In all cases, the measurement gives the value m_0, which is the mass of the body in the aether frame.

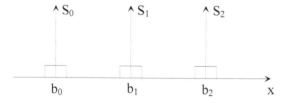

Figure 17. According to relativity, the rest masses of the three bodies are identical. This point of view is not shared by the fundamental aether theory.

Let us now examine the consequences of these results in the following example. Consider three inertial frames S_0, S_1 and S_2, and let three bodies of masses m_0, m_1 and m_2 be at rest in each of the three frames. The masses were initially identical in reference frame S_0 and equal to m_0, before being transported into their respective reference frame. We propose to determine the effect of motion on these masses. (See Figure 17.)

Point of view of the conventional theory of relativity

Measured by an observer at rest with respect to one of the bodies, the mass remains equal to m_0 in all cases. Therefore, for observer S_1, we have

$$m_2^1 = \frac{m_0}{\sqrt{1 - v_{12}^2/C^2}}, \quad (8.7)$$

where m_2^1 designates the relativistic mass of body b_2 as measured by observer S_1, and v_{12} designates the relative speed of reference frames S_1 and S_2.

If we suppose that $v_{12} \ll C$, expression (8.7) can be written to first order, as follows

$$m_2^1 \cong m_0\left(1 + \frac{1}{2}v_{12}^2/C^2\right). \quad (8.8)$$

So that, viewed by observer S_1, the energy of body b_2 is

$$m_2^1 C^2 \cong m_0 C^2 + \frac{1}{2}m_0 v_{12}^2.$$

(This corresponds to the sum of the rest energy and the kinetic energy needed by b_2 to pass from S_1 to S_2.)

For an observer at rest in reference frame S_0, the energy of b_2 is different. Designating m_2^0 the mass of body b_2 as measured by the observer in S_0, we have, (for $v_{02} \ll C$):

$$m_2^0 C^2 \cong m_0 C^2 + \frac{1}{2}m_0 v_{02}^2,$$

and the energy of body b_1 is assumed to be

$$m_1^0 C^2 \cong m_0 C^2 + \frac{1}{2}m_0 v_{01}^2,$$

so that, for observer S_0 the kinetic energy needed by the body b_2 to move from S_1 to S_2 is

$$\left(m_2^0 - m_1^0\right)C^2 \cong \frac{1}{2}m_0\left(v_{02}^2 - v_{01}^2\right).$$

This result is different from the measurement made by observer S_1, $m_0 v_{12}^2/2$, although, obviously, it should be the same. This is a serious internal contradiction that affects special relativity.

Viewpoint of the fundamental aether theory

In our book *Relativité et Substratum Cosmique*,[3] the results below were seen as a stumbling block for aether theories, because they lead to an expression for kinetic energy different from the usual expression. Nevertheless, objections to the relativity principle and present-day arguments in favour of the aether and of the anisotropy of the speed of light compel us to reassess our earlier point of view.

We now go back to the figure with the three bodies, and suppose that S_0 is the fundamental inertial frame, and S_1 and S_2 two inertial frames aligned with S_0. According to the fundamental aether theory, m_2^0 and m_2^1 have no meaning. A body at rest in a given inertial frame has only one real mass. The mass of the body b_2 is:

$$m_2 = \frac{m_0}{\sqrt{1 - v_{02}^2/C^2}}, \tag{8.9}$$

and the mass of b_1:

$$m_1 = \frac{m_0}{\sqrt{1 - v_{01}^2/C^2}}. \tag{8.10}$$

Conversely, as we will see, the rest mass of a body will not be m_0 in the different inertial frames. From (8.9) and (8.10) we obtain

$$m_2 = m_1 \frac{\sqrt{1 - v_{01}^2/C^2}}{\sqrt{1 - v_{02}^2/C^2}}. \tag{8.11}$$

If we now suppose that $v_{02} \ll C$, m_2 reduces to

$$\begin{aligned} m_2 &\cong m_1 + \frac{m_1}{2C^2}\left(v_{02}^2 - v_{01}^2\right) \\ &\cong m_1 + \frac{m_1}{2C^2}\left(v_{12}^2 + 2v_{01}v_{12}\right). \end{aligned} \tag{8.12}$$

This expression is different from (8.8). It contains a term depending on v_{01} which vanishes when S_1 is at rest with respect to S_0. We also see that expression (8.11), which connects any pair of inertial frames, assumes a mathematical form different from (8.9) and (8.10). These results are incompatible with the relativity principle.

We also note that, when $v_{12} \rightarrow 0$ or in other words when $v_{02} \rightarrow v_{01}$, the terms depending on v_{01} and v_{02} in expression (8.12) cancel. Thus, m_1 represents the rest mass assumed by the aforementioned bodies when they are at rest in reference frame S_1. This is a different result from special relativity for which the rest mass is m_0 in any inertial frame.

Nevertheless, we must distinguish the absolute rest mass m_0 from the other rest masses measured in inertial frames that are in motion with respect to the aether frame. Note, however, that when $v_{12} \gg v_{01}$, and $v_{01} \ll C$, expression (8.11) reduces to

$$m_2 \cong \frac{m_1}{\sqrt{1 - v_{02}^2/C^2}} \cong \frac{m_1}{\sqrt{1 - v_{12}^2/C^2}},$$

and since $m_1 \cong m_0$, we obtain

$$m_2 \cong \frac{m_0}{\sqrt{1 - v_{12}^2/C^2}}.$$

This applies, for example, to particles moving at high speed with respect to the Earth frame, while the Earth moves at low speed with respect to the aether frame ($\cong 300$ km/sec.). In such cases, the Earth can be regarded as almost at rest with respect to the Cosmic Substratum. So the relativistic approach and the fundamental approach lead to practically equivalent results.

The question of reciprocity

This question makes a crucial distinction between relativity and fundamental aether theories. According to relativity, when a body is transported from one inertial system S_0 to another S_1, viewed from S_0, its mass is supposed to be

$$m_1 = \frac{m_0}{\sqrt{1 - v_{01}^2/C^2}}.$$

But conversely, if the mass comes back to S_0, viewed from S_1 it will also appear equal to

$$m_1 = \frac{m_0}{\sqrt{1 - v_{01}^2/C^2}}.$$

For the treatment in the fundamental aether theory, let us assume that S_0 is the fundamental frame. If the mass is at rest in frame S_1, we also have

$$m_1 = \frac{m_0}{\sqrt{1 - v_{01}^2/C^2}},$$

where $m_1 > m_0$. Indeed we have been compelled to supply energy to the body in order to move from S_0 to S_1; but if the mass returns to S_0, the energy is restored. All observers (including the one in frame S_1) will conclude that the real mass is equal to

$$m_0 = m_1 \sqrt{1 - v_{01}^2/C^2}.$$

This result is in total contradiction with relativity, but it is the only one in agreement with mass-energy conservation.

Important remark

In the fundamental aether theory, we must distinguish the total available energy of a body (which is equal to the sum of the rest energy m_0C^2 and the kinetic energy with respect to the fundamental frame), from the available energy of the body with respect to any other inertial frame, which is smaller than the previous energy, and takes another mathematical form.

In the example discussed earlier, the total available energy of body b_2 is

$$m_2 C^2 = m_0 C^2 \left(1 + \frac{1}{2}v_{02}^2/C^2\right) + \text{small terms of higher order.}$$

(This notion has no equivalent in conventional relativity for which the energy of a body is entirely relative and depends on its speed with respect to another body.)

And the available energy of body b_2 with respect to frame S_1 is

$$m_2 C^2 - \frac{1}{2}m_0 v_{01}^2 = m_0 C^2 + \frac{1}{2}m_0\left(v_{02}^2 - v_{01}^2\right).$$

Possible measurement of the absolute speed of an inertial system

Assuming that $v_{02} \ll C$, the kinetic energy needed to pass from S_1 to S_2 reduces to:

$$\Delta E_C \cong \frac{1}{2}m_0\left(v_{12}^2 + 2v_{01}v_{12}\right).$$

Knowing ΔE_C and v_{12}, it is theoretically possible to measure the absolute speed, v_{01}, of the inertial system S_1, that is:

$$v_{01} \cong \frac{\Delta E_C - \frac{1}{2}m_0 v_{12}^2}{m_0 v_{12}}.$$

This result is also in contradiction with the relativity principle.

Conservation of energy

In our opinion, the mass-energy conservation law is beyond question and should apply exactly in any inertial frame. Nevertheless, at high speeds, the role of the aether wind would not be completely negligible and should be taken into account in any event where an exchange of energy occurs.

Chapter 9

Synchronization Procedures and Light Velocity

Measuring the speed of light with one or two clocks by the Einstein-Poincaré procedure

In order to measure the speed of light, we can use one or two clocks. When we use one clock, the signal is sent from the clock toward a mirror, and, after reflection, comes back to its initial position. In this case, what we measure is the *apparent* average round trip velocity of the light signal.

As we saw in Chapter 6 (formula (6.10)), even if we subscribe to the Lorentz assumptions, which assume the anisotropy of the one-way speed of light in the Earth frame, the theory demonstrates that this average velocity is (erroneously) found equal to C in any direction of space. It also appears independent of the relative speed of the frame in which it is measured with respect to the aether frame. (These results follow from the systematic measurement errors already mentioned.)

Therefore, *a priori*, the use of two clocks seems justified in order to accurately measure the one-way speed of light. With this aim in view, we need first to synchronize two distant clocks A and B.

In the Einstein-Poincaré procedure, this requires two steps. First, we send a light signal from clock A to clock B at instant t_0; after reflection the signal comes back to A at instant t_1. Then we send another signal at instant t'_0. The clocks will be considered synchronous if, when the signal reaches clock B, the display of clock B is:

$$t'_0 + \frac{t_1 - t_0}{2} = t'_0 + \varepsilon,$$

where ε is the *'apparent'* average transit time of the signal measured with the retarded clocks attached to the Earth frame; but in the Einstein-Poincaré procedure it is identified with the one way transit time of light.

As we saw in Chapter 6 (formula (6.10)), $\varepsilon = \ell_0/C$ in any direction of space; and since the distance AB is always found equal to ℓ_0, the speed of

From Galileo to Lorentz... and Beyond
Joseph Lévy (Montreal: Apeiron 2003)

light is found equal to C in the same way as when we use one clock. Thus, even though the speed of light is given by formulas (6.5) and (6.6), the Einstein-Poincaré procedure finds C.

It would therefore seem justified to test another method, *i.e.*, the slow clock transport procedure.

Measuring the speed of light by the slow clock transport procedure

Many physicists believe that an exact measurement of the speed of light can be obtained by the slow clock transport method. The procedure consists of synchronizing two clocks A and B at a point O' in the Earth frame, and then transporting clock B to a distance from A at low speed ($v \ll C$.) The problem has been envisaged in different ways by different authors.[56-64]

A priori, it would appear that, since the transport is very slow and $v \to 0$, the motion would have no perceptible influence on the time displayed by clock B, and that the two clocks would remain almost synchronized all the time. But is this really the case?

The special relativity viewpoint

If we regard the assumptions of special relativity as indisputable, then absolute speeds have no meaning: only relative speeds exist. On this basis, clock B will display[*]:

$$t' = t\sqrt{1 - v^2/C^2} \cong t\left(1 - \frac{1}{2}\frac{v^2}{C^2}\right),$$

where t is the display of clock A. (Note that for convenience we have supposed that the two clocks display $t_0 = 0$ at the initial instant.)

Once clock B has stopped (at point P) its lag behind clock A will remain constant. The synchronism discrepancy between clocks A and B is then, to first order

$$\Delta t = \frac{1}{2}\frac{v^2}{C^2}T,$$

where T designates the time displayed by clock A when clock B reaches point P.

The speed of light will thus appear to be:

$$\frac{O'P}{T - \Delta t} \cong \frac{O'P}{T} + \frac{O'P\Delta t}{T^2}. \qquad (9.1)$$

Since $v \to 0$, expression (9.1) reduces to

---

[*] Note that the value of the speed of light is assumed to be known. The measurement therefore consists in verifying whether the results obtained by this method are in agreement with the premises.

$$\frac{O'P}{T}.$$

The experimental value of the speed of light obtained is C.

Since the measurements of $O'P, T$ and Δt are assumed to be exact, special relativity concludes that the real value of the speed of light in the Earth frame is C. Hence, if we admit the assumptions of special relativity, the speed of light measured by the method of slow clock transport is in agreement with the assumed hypotheses. But the method in no way justifies these hypotheses.

Viewpoint of the fundamental aether theory

It is interesting to see whether the above results can be obtained with basic hypotheses different from those of special relativity. Today, we have strong arguments in favour of the Lorentz assumptions. Several of them have been reviewed in the earlier chapters of this book. According to Lorentz, the speed of light is C exclusively in the aether frame.

Notice that, if absolute speeds are taken into consideration, then there is no real slow clock transport, since the absolute motion of the Earth is added to that of the transported clock. And various arguments, already presented in the previous chapters, demonstrate that the motion of the Earth cannot be ignored.

If the method is reliable, it should give a value for the one-way speed of light in agreement with the assumed hypotheses. We will now verify this point. Two cases will be considered.

The light ray travels in the direction of motion of the Earth frame with respect to the aether frame

Consider two inertial systems of co-ordinates S_0 and S_1. S_0 is at rest in the Cosmic Substratum, and S_1 is firmly linked to the Earth frame. Initially the two frames are coincident. At this instant, a vehicle equipped with a clock starts from the common origin and moves slowly and uniformly along the x-axis of frame S_1 toward a point P in this frame. We suppose that the x-axis is aligned along the direction of motion of the Earth with respect to the Comic Substratum. (See Figure 18.) v_{01} is the speed of the Earth with respect to the fundamental frame S_0, v_{02} is the speed of the vehicle with re-

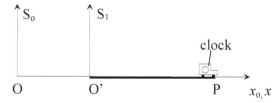

Figure 18. Synchronization of two clocks placed at O' and P by the slow clock transport method.

spect to S_0, and v_{12} the speed of the vehicle with respect to S_1.
(Note that, for a short time, the motion of the Earth with respect to the Cosmic Substratum can be considered rectilinear and uniform. If this were not the case, the bodies standing on the Earth platform would be submitted to perceptible accelerations.)

The duration of the transport should be short enough so that the orbital and rotational motions of the Earth would not significantly affect the measurement.

When the vehicle reaches point P, it stops. The real time needed to reach point P is given by

$$t_r = \frac{\ell}{v_{02} - v_{01}} = \frac{\ell_0 \sqrt{1 - v_{01}^2/C^2}}{v_{02} - v_{01}},$$

where ℓ is the length of O'P (which is contracted because of the motion of the Earth with respect to the Cosmic Substratum), ℓ_0 is the length that O'P would assume if it were at rest in the aether frame, t_r is the vehicle's real transit time from O' to P. It is the time that a clock attached to the aether frame, opposite the vehicle at the instant when it reaches point P, would display. (Recall that, in the fundamental aether theory, real speeds obey the Galilean law of composition of velocities.)

But the clock in the vehicle (B) is slow relative to the clock in frame S_0, and will display the time

$$\frac{\ell_0 \sqrt{1 - v_{01}^2/C^2} \sqrt{1 - v_{02}^2/C^2}}{v_{02} - v_{01}}.$$

Now the clock placed at the origin O' of the Earth system (A) slows down relative to a clock attached to frame S_0 opposite it. When the vehicle reaches point P, it will display the time:

$$\frac{\ell_0 \sqrt{1 - v_{01}^2/C^2} \sqrt{1 - v_{01}^2/C^2}}{v_{02} - v_{01}}.$$

(This implies that, for an instantaneous event occurring at point P, all the clocks attached to the Cosmic Substratum display the same time.)

Thus, between clock B and clock A, we find a synchronism discrepancy equal to:

$$\frac{\ell_0 \sqrt{1 - v_{01}^2/C^2}}{v_{02} - v_{01}} \left(\sqrt{1 - v_{01}^2/C^2} - \sqrt{1 - v_{02}^2/C^2} \right)$$

$$\cong \frac{\ell_0 \sqrt{1 - v_{01}^2/C^2}}{v_{02} - v_{01}} \left(1 - \frac{1}{2} v_{01}^2/C^2 - 1 + \frac{1}{2} v_{02}^2/C^2 \right) \quad (9.2)$$

$$\cong \frac{\ell_0}{2C^2} \sqrt{1 - v_{01}^2/C^2} \, (v_{02} + v_{01}).$$

We can see that, once the vehicle has stopped, the discrepancy will remain constant.

Speed of light along O'P

If we assume the Lorentz postulates, the real time of light transit along the distance ℓ is theoretically

$$\ell_0 \frac{\sqrt{1-v_{01}^2/C^2}}{C-v_{01}}.$$

We suppose here, *a priori*, that the speed of light with respect to frame S_1 is $C - v_{01}$. This is intentional, since we want to check whether the results are in agreement with the premises.

Now, as a result of clock retardation, (and without making allowance for lack of synchronism) the display of a clock in frame S_1 placed at point P when the signal reaches this point should be:

$$\frac{\ell_0 \sqrt{1-v_{01}^2/C^2}\sqrt{1-v_{01}^2/C^2}}{C-v_{01}} = \frac{\ell_0}{C-v_{01}}\left(1-v_{01}^2/C^2\right).$$

If, in addition, we take into account the synchronism discrepancy given by formula (9.2), the apparent (measured) light transit time will be:

$$\frac{\ell_0}{C-v_{01}}\left(1-v_{01}^2/C^2\right) - \frac{\ell_0}{2C^2}\sqrt{1-v_{01}^2/C^2}\left(v_{02}+v_{01}\right). \qquad (9.3)$$

Ignoring the terms of high order, expression (9.3) reduces to

$$\frac{\ell_0}{C}\left(1+\frac{v_{01}-v_{02}}{2C}\right) = \frac{\ell_0}{C}\left(1-\frac{v_{12}}{2C}\right).$$

Now, since the measured length of O'P is always found equal to ℓ_0, the *apparent* speed of light measured with the transported clock will be

$$\frac{\ell_0}{\frac{\ell_0}{C}\left(1-\frac{v_{12}}{2C}\right)} = \frac{C}{1-\frac{v_{12}}{2C}} \cong C\left(1+\frac{v_{12}}{2C}\right) = C + \frac{v_{12}}{2}.$$

Since v_{12} is taken as small as possible, the apparent speed of light is found equal to C. Therefore, even if the real speed of light is $C - v_{01}$, the slow clock transport method will (erroneously) find C in the same way as the Einstein-Poincaré method.

Therefore the two methods can be considered equivalent.

General case

We now measure the speed of light along a rod O'B making an angle θ with the *x*-axis of a system of co-ordinates S_1, firmly tied to the Earth frame. (See Figure 19.) Note that the *x*-axis of S_1 is aligned along the direction of motion of the Earth with respect to the Cosmic Substratum. For a short period of time this motion can be considered rectilinear and uniform.

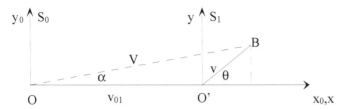

Figure 19. Synchronization of two clocks placed at O' and B by the slow clock transport method.

(Note also that the rod is in the x, y plane, but obviously, provided θ remains the same, the following reasoning would be identical in any plane passing by the x_0, x-axis.)

We can choose a system of co-ordinates S_0 in the Cosmic Substratum such that S_0 and S_1 are initially coincident. At this instant, a vehicle leaves the common origin, and moves slowly and uniformly along the rod toward point B.

As we saw in Chapter 6, formula (6.8), due to length contraction along the x_0, x-axis, the length of the rod is given by

$$\ell = \frac{\ell_0 \sqrt{1 - v_{01}^2/C^2}}{\sqrt{1 - v_{01}^2 \sin^2 \theta/C^2}},$$

where v_{01} is the speed of the Earth with respect to the fundamental frame S_0. We will designate as v the real speed of the vehicle with respect to S_1, and V its real speed with respect to S_0. (See Figure 19.)

The real time needed by the vehicle to reach point B is ℓ/v, but the apparent time in frame S_1, after allowance is made for clock retardation, is

$$\frac{\ell}{v} \sqrt{1 - v_{01}^2/C^2}.$$

The apparent time as measured with a clock inside the vehicle is

$$\frac{\ell}{v} \sqrt{1 - V^2/C^2}.$$

As a result, the synchronism discrepancy between the apparent time displayed by a clock attached to frame S_1 placed at point O' and the clock in the vehicle is

$$\Delta = \frac{\ell}{v} \left(\sqrt{1 - v_{01}^2/C^2} - \sqrt{1 - V^2/C^2} \right).$$

We easily find that

$$V^2 = v^2 \sin^2 \theta + \left(v_{01} + v \cos \theta \right)^2,$$

and

$$V = v_{01}\sqrt{\frac{v^2}{v_{01}^2}+1+\frac{2v}{v_{01}}\cos\theta}. \tag{9.4}$$

If the inequality $v \ll v_{01}$ is taken into account, expression (9.4) reduces to

$$V \cong v_{01}\left(1+\frac{v}{v_{01}}\cos\theta\right) = v_{01} + v\cos\theta. \text{ (See Figure 19.)}$$

Therefore, to first order, Δ becomes

$$\frac{\ell}{C^2}\left(v_{01}\cos\theta + \frac{1}{2}v\cos^2\theta\right).$$

Measuring the speed of light along O'B by the slow clock transport procedure

Let us now suppose that we place, in O' and B, two clocks that have been (apparently) synchronized by the slow clock transport method. In fact, there is a synchronization error equal to Δ.

The real speed of light along the rod from O' to B is (as seen in Chapter 6, formula (6.5)).

$$C_1 = -v_{01}\cos\theta + \sqrt{C^2 - v_{01}^2\sin^2\theta}.$$

As a result of clock retardation, but without the synchronism discrepancy effect, the *apparent* time needed by the light ray to reach point B should be

$$T_L = \frac{\ell}{C_1}\sqrt{1-v_{01}^2/C^2}.$$

However, we must allow for the synchronism discrepancy, so that the 'apparent' (measured) transit time of light will be:

$$\frac{\ell}{C_1}\sqrt{1-v_{01}^2/C^2} - \Delta. \tag{9.5}$$

Ignoring terms of high order, expression (9.5) reduces to

$$\ell_0 \frac{\left(1-\frac{v_{01}}{C}\cos\theta - \frac{1}{2}\frac{v}{C}\cos^2\theta\right)}{C\left(1-\frac{v_{01}}{C}\cos\theta\right)}.$$

Since the rod O'B is measured with a contracted meter stick, it appears equal to ℓ_0.

The apparent speed of light is then

$$C_{app} = \frac{\ell_0}{T_L - \Delta} = \frac{C\left(1-\frac{v_{01}}{C}\cos\theta\right)}{1-\frac{v_{01}}{C}\cos\theta - \frac{1}{2}\frac{v}{C}\cos^2\theta}.$$

Since $v \to 0$, $C_{app} \to C$, which is different from its real value C_1, as shown earlier.

Therefore, contrary to what is often claimed,[65] the slow clock transport method does not allow an exact measurement of the speed of light. It is approximately equivalent to the Einstein-Poincaré method, and like this method, gives the erroneous value C for all measurements.

It is interesting to note that, even though the speed of light is not constant, it is found constant when standard methods of synchronization are used. Consequently, these methods must be seen as inadequate. Nevertheless, although it is difficult to measure the speed of light, it is not in principle impossible, as we saw in the section "The relativity principle" in Chapter 2.

References

1. J. Levy, "Relativity and Cosmic Substratum," in: *Physical Interpretations of Relativity Theory* (P.I.R.T.), Imperial College, London, 6-9 September, 1996, p 231, Ed M.C Duffy, E-mail Michael.duffy9@btopenworld.com. A symposium sponsored by the British Society for the Philosophy of Science.
2. *ibid.*, "Some important questions regarding Lorentz-Poincaré's theory and Einstein's relativity I," in: *Physical Interpretations of Relativity Theory* (P.I.R.T.), 1996. Late papers, p 158.
3. *ibid., Relativité et Substratum Cosmique* 230 p., Ed Lavoisier, Cachan, France, (1996). E-mail edition@Lavoisier.fr, Tel 01 47 40 67 00.
4. *ibid.*, "Some important questions regarding Lorentz-Poincaré's theory and Einstein's relativity II," *Physical Interpretations of Relativity Theory* (P.I.R.T.), 1996. Supplementary papers, p 178.
5. *ibid.*, "Is the relativity principle an unquestionable concept of physics," *Physical Interpretations of Relativity Theory* (P.I.R.T.), 1998, late papers p 156 and "Hidden variables in Lorentz transformations," (P.I.R.T.), 1998. Supplementary papers p 86.
6. *ibid.*, "Is simultaneity relative or absolute" in *Open questions in relativistic physics,* Ed F. Selleri, (Apeiron, 4405 rue St Dominique, Montreal, Quebec, H2W 2B2, Canada, E-mail apeiron@vif.com.)
7. *ibid.*, "Invariance of light speed reality or fiction," *Physics Essays* 6 (1993) p 241.
8. H.A. Lorentz, "Electric phenomena in a system moving with any velocity less than that of light," in *The principle of relativity* (Collection of original papers on relativity, New York, Dover, 1952.)
   G. Galeczki, *Physics Essays* 8 (1995) p 591, see note 2.
9. O. Heaviside, *Phil. mag.* 27 (1889) p 324.
10. - G.F. Fitzgerald, *Science*, 13 (1889) p 390.
11. H.A. Lorentz, *Zitt d Akad, v wet te Amsterdam* (1892-93), p 74.
12. J. Larmor, *Phil Trans Roy Soc* London (1897) p 190-205.
    Aether and Matter, (Cambridge Univ. Press, 1900.)
13. J.S. Bell, *Physics World* (Sept 1992) p 31-35.
14. S.J. Prokhovnik, *Z. Naturforsch.*, 48a, 925 (1993) p 928.
15. S. Marinov, *Spec. Sci. Tech* 3 (1980a) p 57, *The thorny way of truth* (East. West, Graz, Austria, 1984,) *Gen. Rel. Grav* 12 (1980) p 57.
16. H.A. Lorentz, *Proc Acad Sci Amsterdam*, 6, (1904) and *The principle of Relativity*, (New-York, Dover 1952.) p 11.
17. W. Voigt, *Nackrichten von der K.G. d. N. zu Göttingen* 2 (1887) p 41.
18. H. Poincaré "Sur la dynamique de l'électron" and "Lecture given in Lille 1909" in *La mécanique nouvelle* (Ed Jacques Gabay, Sceaux, 1989.)
19. A. Einstein, *Annalen der Physik* 17 (1905) p 891.
    The Principle of Relativity, (New- York, Dover, 1952.)
20. H. Minkowski, "Space and Time," in *The Principle of Relativity*, (New York, Dover, 1952.)
21. H. Poincaré, *La science et l'hypothèse* (Paris, Champs Flammarion, 1968.)
22. S.J. Prokhovnik, 1-*The logic of special relativity* (Cambridge University press, 1967.)
    2-*Light in Einstein's Universe* (Reidel, Dordrecht, 1985.) References to the articles of G. Builder.
23. W. De Sitter, *Phys Zeitschr*, 14 (1913) p 429.

24 J. Levy. "Is the invariance of the speed of light compatible with quantum mechanics, some new arguments," in: *Advances in Fundamental physics*," Ed M. Barone and F. Selleri (Hadronic Press, Palm Harbor, Fl, USA, 1993.)
25 D. Sadeh, *Phys Rev Lett* 10 (1963) p 271.
26 T. Alväger et al., *Phys Lett* 12 (1964) p 260.
27 A. Einstein, *La relativité* (Payot, Paris.)
28 F. Selleri, Inertial systems and the transformations of space and time, *Physics Essays*, 8 (1995) p 342.
29 R. Mansouri and R.U. Sexl, *General relativity and gravitation* 8 (1977) p 497.
30 J. P Wesley, "Evidence for Newtonian absolute space and time," in: *Open questions in relativistic physics*, p 255, Ed. F. Selleri (Apeiron 4405, St Dominique, Montreal, Quebec, H2W 2B2, Canada, E-mail apeiron@vif.com). See in particular Table 1 and section 8.
    H. E. Wilhelm, *Physics Essays* 6 (1993) p 420.
31 ibid, *Selected topics in Advanced fundamental physics*, (Ed. Benjamin Wesley, Blumberg, 7712, Germany, 1991.)
32 F. Tangherlini, *Suppl Nuovo Cimento*, 20 (1961) p 1.
33 T.E. Phipps Jr, *Found Phys*, 10 (1980) p 289 and *Apeiron*, vol 4 N°2-3, April, July 1997, p 91.
34 P. Cornille, *Physics Essays*, 5 (1992) 262 p 271.
35 G. Galeczki, *Physics Essays*, 8 (1995) p 591.
36 J.S.Bell, "How to teach special relativity," in *Speakable and unspeakable in Quantum mechanics,* (Cambridge University press, 1987) and George Francis Fitzgerald, *Physics World*, September 1992.
37 F. Selleri, "Le principe de relativité et la nature du temps," *Fusion* 66, mai juin 97, p 50, Paris. Remarks on the transformations of space and time *Apeiron* 4 (1997) p 100-103.
    "On the meaning of special relativity if a fundamental frame exists," in *Progress in new cosmologies*, p 269-284, Ed H. Arp et al., Plenum, New York London, 1993.
38 R.H. Dishington, Selleri's Theorem, *Apeiron* 5, 3-4 (1998) p 239.
39 Lord Rayleigh, *Phil Mag* 4 (1902) p 678.
40 D.B. Brace, *Phil Mag* 7 (1904) p 317.
41 F.T. Trouton and H.R. Noble, *Proc. Roy Soc* 72 (1903) p 132 and *Phil Trans* A 2 (1903) p 167.
42 F.T. Trouton and A.D Rankine, On the electrical resistance of moving matter, *Proc Roy Soc* 80 (1908) p 420.
43 C.T. Chase, *Phys Rev* 30 (1927) p 516.
44 R. Tomashek, *Ann. d. Physik* 73 (1924) p 105, 78 (1925) p 743, 80 (1926) p 509, 84 (1927) p 161.
45 A.B. Wood, G.A. Tomlison and L. Essex. The effect of the Fitzgerald-Lorentz contraction on the frequency of longitudinal vibration of a rod, *Proc.Roy.Soc* 158 (1937) p 6061.
46 M.A. Tonnelat, *Histoire du principe de relativité* (Flammarion Paris p 97-107).
47 C.W. Sherwyn, *Phys Rev A* 35 (1987) p 3650.
48 R. Anderson, I. Vetharaniam, G.E Stedman, *Physics Reports* 295, 93-180 (1998) p108.
49 P. Beckmann, *Einstein Plus Two* (The Golem press, Boulder, Co, 1987).
50 D. Mitsopoulos, *Physics Essays* 6 (1993) p 233.
51 V.I. Makarov, L'aberration astronomique, private communication.
52 O. Lodge, Aberration problems, *Phil Trans Roy Soc*, London, A 184 (1893) p 727.
53 J.P. Wesley, "My memories of Stefan Marinov" in *Physics as a Science*, Ed. G. Galeczki and P. Marquardt (Hadronic Press, Palm Harbor, FL, USA 1998) p 29.
    J.P. Wesley, private communication.
    J.J Smulsky, In memory of Stefan Marinov, *Apeiron* 5, 1-2 (1998.)
54 F. Rohrlich, *Am J. Phys* 58 (4) (1990) p 348.
55 G. N. Lewis, *Phil Mag* 16 (1908) p 705.
56 A.S. Eddington, *The mathematical theory of relativity*, 2nd ed. (Cambridge University Press, Cambridge 1924)
57 H. Reichenbach, *The philosophy of space and time* (Dover, New York 1958)

58 A. Grünbaum, *Philosophical problems of space and time* (A. Knopf, New York, 1963)
59 P. W. Bridgman, *A Sophisticate's primer of relativity* (Wesleyan University Press, Middletown, 1962)
60 B. Ellis and P. Bowman, Conventionality in distant simultaneity, *Phil, Sci* 34 (1967) p 116-136.
61 A. Grünbaum, Simultaneity by slow clock transport in the special theory of relativity, *Phil Sci*, 36 (1969) p 5-43.
62 Yu. B. Molchanov "On a permissible definition of simultaneity by slow clock transport," in: *Russian Einstein Studies*, Nauka, Moskow, 1972.
63 J. A. Winnie Special relativity without one-way velocity assumptions *Phil sci*, 37, p 81-89 and p 223-238 (1970).
64 R.G. Zaripov, Convention in defining simultaneity by slow clock transport," *Galilean Electrodynamics* 10, May June 1999 p 57.
65 R. Anderson *et al.*, *Physics reports* 295 (1998) p 93-180. See in particular p 100, where the authors criticize attempts to measure the one-way speed of light by means of the slow clock transport procedure. References to Krisher *et al.*, Nelson *et al.*, Will, Haughan *et al.* and Vessot.
66 J. Levy, "Extended space-time transformations derived from Galilei's," in: *Physical Interpretations of Relativity Theory* (P.I.R.T.), 15-18 September, 2000, p 211. Imperial College, London. (An updated version will be published.)

Further references with comments

M. Allais, *L'anisotropie de l'espace* (Clement Juglar, Paris 1997) 750 pages.

> The author comments on experiments he performed with a paraconic pendulum which, according to him, demonstrated the anisotropy of space. He also refers to the experiments of Miller and Esclangon, which lead to identical conclusions.
>
> Miller repeated the experiments of Michelson and Morley a number of times and concluded: "...Since the theory of relativity postulates an exact null effect from the aether drift experiment which had never been obtained in fact, the writer felt impelled to repeat the experiment in order to secure a definitive result." (*L'anisotropie de l'espace*, p 383.)
>
> Allais notes (p. 405 of his book note 9 and p. 581 note 2) that "it is erroneous to repeat that the experiments of Michelson and Morley of 1887 gave a completely negative result since they showed a fringe shift corresponding to a speed of 8 km/sec."
>
> *Comments*
>
> 1. The author does not quote the modern Michelson-Morley type experiments, the sensitivity of which has been increased considerably. For example, Joos (1930) detected a difference in the two-way speed of light in two perpendicular directions of 1.5 km/sec. Jaseja *et al.* (1964), using masers mounted at right angles, found 1 km/sec. Brillet and Hall in 1979 performed an experiment of very high sensitivity (about 30×10^{-5}) which showed a two-way speed of light almost identical in two perpendicular directions. The difference was of the order of 16 m/sec. Therefore, contrary to the author's claims, modern experiments tend to confirm the apparent isotropy of the two-way speed of light.

2. The experiments of Miller quoted by Allais are not concerned with the anisotropy of the one-way speed of light, which amounts to approximately 300 km/sec: they only measure the *apparent* two-way speed of light.

V. Bashkov, *Geometrization of Physics*, University of Kazan, Russia.

V Bashkov was the organizer of a conference held every two years in Kazan state University, Russia, and author of several fundamental articles on special and general relativity, some in English and others in Russian.

A. Brillet and J.L Hall, *Phys Rev Lett* 42 (1979) p 549.

According to Hayden it is "by far the best Michelson-Morley experiment performed to date. It has been designed to be clear in its interpretation and free of spurious effects."

The authors have handled their data in such a manner that effects that may arise from the Earth's rotation are ignored. The results show a difference of the two-way speed of light in two perpendicular directions which does not exceed 16m/s (8km/s for Michelson and Morley).

G. Builder, *Aust J Phys* 11 (1958a) 279 and 11 (1958b) p 457 and *Philosophy Sci* 26 (1959) p 135.

These articles are historically important, they develop an original viewpoint regarding relativity theory. The ideas of Builder have been presented and developed by his disciple Simon Prokhovnik in different articles and in two reference books. (See Prokhovnik reference.)

G. Cavalleri and C. Bernasconi, *Nuovo Cimento B*, Vol. 104B, Ser. 2, No. 5 (1989) p 545.

The authors propose that, contrary to what is generally believed, light speed invariance and non conservation of simultaneity are not peculiar properties of Einstein's special relativity.

"Simultaneity conservation (or not) and light speed invariance in both Galilean physics and special relativity, depend on the arbitrary way in which the clocks of reference systems are synchronized."

M.C. Duffy, "Aether, cosmology and general relativity," and "The aether, quantum mechanics and models of matter," *Gdansk conference,* Sept. 1995. An extended version of the first article was published in the supplementary papers of the 1998 P.I.R.T. conference, *Physical Interpretations of Relativity Theory,* Imperial College London, 11-14 September 1998 p 16.

The first of these papers attempts a review of the relativistic world aether theories and seeks to identify the links between them. The aether concepts developed by Clube, Cavalleri, Borneas, Wheeler, Nesteruk, Prokhovnik, Ives, and Einstein, among others are analyzed.

The author concludes that a promising role for the aether within general relativity and cosmology has been convincingly demonstrated in recent years.

The second paper points out the role of the aether in the formulation of a grand comprehensive theory unifying relativity and quantum mechanics.

References to Einstein, Dirac, Borneas, Podhala, Jennison, Winterberg, Cavalleri, Eddington, among others.

H.C. Hayden, Yes clocks run slowly, but is time dilated? *Galilean Electrodynamics* 2 (1991) p. 63.

> The author claims that all the experiments intended to prove time dilation have actually demonstrated the slowing down of moving clocks due to their motion through the gravitational field. In effect, no data at all showed any evidence of the symmetry that is central to special relativity.

H.C. Hayden, Is the velocity of light isotropic in the frame of the rotating Earth? *Physics Essays* 4 (1991) p 361.

> The author analyses several experiments purporting to demonstrate Einstein's second postulate. The author examines different classical experiments and their modern versions. For example, the Brillet *et al.* experiment, a modern version of Michelson-Morley, shows that the difference in the two-way speed of light in two perpendicular directions does not exceed 16 m/sec. The experiment of Allan *et al.*, using satellites and several ground stations, confirms the Sagnac experiment. References to Jaseja *et al.*, Allan *et al.*, Brillet and Hall, Sagnac, Michelson and Gale.

L. Kostro, Physical interpretation of Albert Einstein's relativistic ether concept, *Physical Interpretations of Relativity Theory* (P.I.R.T.) 9-12 September 1994 p 206.

> The author gives a thorough analysis of the evolution of Einstein's ideas about the aether.
>
> "Until the end of his life Einstein denied the existence of an ether as it was conceived in 19th century physics, in particular Lorentz's ether which was in the first place a privileged reference frame ... because it violated his principle of relativity. Nevertheless, in 1916, Einstein proposed a new conception of the ether ... which does not violate the principle of relativity because the space-time is conceived in it as a material medium *sui generis* that can in no way constitute a frame of reference."
>
> The author points out that in 1918 Lorentz also presented a model of ether at rest with respect to every reference frame, not only with respect to the preferred frame.
>
> As we have seen in this book, these models of ether lead to consequences in contradiction with well established concepts of physics.

P. Marmet, Einstein's theory of relativity versus classical mechanics, (Newtonphysics Books, Gloucester, ON, Canada K1J 7N4).

> The book by Paul Marmet presents an original vision of space-time theory. The author claims that, using conventional wisdom and conventional logic, classical mechanics can explain all the observed phenomena attributed to relativity. According to the author, "contrary to what Einstein did, all his demonstrations are compatible with mass-energy and momentum conservation." Einstein's relativity principle is not needed in his demonstrations.
>
> Einstein's relativity implies new logic which contradicts conventional logic. The author refutes these views and asserts that there is no need to

give up Newtonian logic in order to reproduce all the key results that relativity theory claims.

Although we do not grasp all of the author's assumptions and we differ with his views, we acknowledge that many points raised in the book are interesting and thought-provoking.

M. Mascart, Sur les modifications qu'éprouve la lumière par suite du mouvement de la source lumineuse et du mouvement de l'observateur *(Annales scientifiques de l'Ecole Normale Supérieure)* 2° serie, t I, p 157 -214, and p 364-420.

> The author gives a detailed account of the results of a number of experiments intended to verify the influence of the motion of the Earth on optical phenomena (experiments of Arago, Fresnel, Fizeau, measurement of the rotating power of quartz, double refraction, *etc.*)
>
> The author concludes that the translation of the Earth has no influence on these optical phenomena. As a result, they do not detect the absolute motion of the Earth. Only relative motions can be observed.
>
> The experiments of Mascart have cleared the way for Potier and Veltmann. (See reference to these authors.) However, the conclusions reached by this author cannot be extended to all types of experiments (for example, Marinov's experiments).

D.C. Miller, Experiment on Mount Wilson (1750m), 1921, quoted by M. Allais in *L'anisotropie de l'espace* p 584.

> The author found that the small fringe shift in Michelson's experiment at this altitude was not smaller than at sea level: in fact, it was a little larger. The anisotropy of the two-way speed of light in two perpendicular directions on Mount Wilson was found to be 10 km/sec, while at sea level it was found to be 6 km/sec.
>
> The most probable conclusion is that these experiments are outdated in comparison to modern Michelson Experiments. (See Brillet and Hall and Hayden references.)

T.E. Phipps Jr, Potier's principle, a trap for aetherists and others, *Galilean Electrodynamics* 3 (1992) p 56.

> The author points out that Potier's principle denies the theoretical possibility of simple optical test of the existence of an aether wind (to first order).
>
> Nevertheless the principle does not apply to all types of experiments, as demonstrated in Chapter 6 of this book.

T.E. Phipps Jr, Absolute simultaneity with and without light signals, *Galilean Electrodynamics* (May, June 1996) p 43.

> The author bases his views on neo-Hertzian electromagnetism, which is Galilean invariant, rather than on Maxwellian electromagnetism. According to Phipps, if signals are described by neo-Hertzian equations, then absolute simultaneity appears compatible with a relativity principle and with length invariance.

Y. Pierseaux, La "Structure fine de la relativité resteinte (Editions l'Harmattan, Paris).

> An exhaustive comparative study of the orthodox approaches of Poincaré and Einstein to relativity. The author demonstrates that, contrary to the opinion of Whittaker and others, the two theories differ in many respects: primacy of continuous processes, existence of aether, hidden variables in Poincaré, primacy of discontinuity, non existence of aether and absence of hidden variables in 1905 Einstein's approach. The book provides the reader with a host of quotations and references.

M.G. Sagnac, M.E. Bouty. L'ether lumineux démontré par l'effet du vent relatif d'ether dans un interferometre en rotation uniforme" and "Sur la preuve de la réalité de l'ether lumineux par l'expérience de l'interferographe tournant, *Comptes-rendus Acad Sci* Paris, vol 157, (1913) p 708 and 1410.

> A non null fringe shift is observed when light is sent in opposite directions around a rotating table. The experiment lends support to the hypothesis of a non entrained aether.

F. Selleri. *Lezioni di Relativita.* (Ed Progedit, Bari, Italy, March 2003).

> These lessons explain in an elementary but critical way the special theory of Relativity and the conceptual foundations of the general theory, giving ample place to the most important ideas and to their philosophical implications. In addition to the orthodox theory, the author presents the most important investigations made during the last ten years. According to him, there exists an infinite number of theories equivalent to "special relativity" all based on the existence of a privileged reference frame. He asserts that certain phenomena break the equivalence, and are better explained using absolute synchronization. A return to Lorentz aether is finally possible. Although our approaches are quite different, the ideas developed in this book show several points of convergence with our views.

W. Veltmann, *Astron Nachr* vol.76 (1870) p 129-144, and A. Potier, *Journal de physique* (Paris) vol 3 (1874) p 201-204.

> Using Fermat's principle the authors propose that it is impossible by means of an optical experiment to observe a first order aether wind in v/C. However, it should be borne in mind that the idea is not applicable to experiments such as Marinov's toothed wheels experiment, or any other experiment described in Chapter 6 of this book.

C. K. Whitney. "How can paradox happen?" in: *Physical Interpretations of Relativity Theory* (P.I.R.T.), 15-18 September 2000, p 338.

> Einstein's relativity leaves us today with a number of paradoxes. The paper develops the view that physical reality is one thing while our conceptual model for it is quite another. When the two do not match, we will make wrong inferences from data which can be inconsistent and lead to apparent paradoxes. The possibility that a wrong physical model may be embedded in Einstein's relativity theory is traced to the sequence of historical development: in the early days of his work, Einstein

worked with Maxwell's electromagnetic theory, but not modern quantum mechanics.

The proposed model includes facts that have appeared since the advent of quantum mechanics. The author asserts that it predicts the main features of special relativity without paradoxes, as well as the main predictions of general relativity.

F. Winterberg. "Derivation of quantum mechanics and Lorentz invariance from Newton's ultimate object conjecture at the Planck length," in: *Physical Interpretations of Relativity Theory* (P.I.R.T.) 1996, Late papers p 322.

The paper is dedicated to S. Prokhovnik and makes use of this author's approach, which, assuming length contraction, demonstrates the isotropy of the two-way transit time of light and then connects clock retardation to length contraction.

As noted in this monograph, we share with F. Winterberg some of Prokhovnik's views, but not all of them. Here is a quote from the author:

"By identifying Newton's ultimate objects with Planck mass particles, but also assuming that there are negative besides positive Planck mass particles, and finally assuming that space is densely filled with an equal number of positive and negative mass particles locally interacting with each other within a Planck length, we can derive quantum mechanics and special relativity from Newtonian mechanics."

Index

absolute motion, 35-6, 56, 60, 65, 79, 90
absolute simultaneity, 12, 90
absolute space, 36, 60, 86
aether, v, 1-4, 8, 12, 13, 15, 19, 20-22, 29-31, 33, 35, 36, 39, 40, 42, 44, 47, 48, 50, 51, 56-58, 60, 61, 63-67, 74, 76, 77, 79, 80, 85, 87, 88, 90, 91
aether frame, 1, 4, 13, 19, 48, 66, 80
aether wind, 47, 51, 76, 90, 91
anisotropy of the one way speed of light, 48, 51, 55, 58, 87
apparent speed of light, 81, 83
apparent time, 45, 82, 83
Aristotle, 59
balance, 12
beta Aurigae, 4
clock display, 4, 12-14, 42, 56, 65, 81
clock retardation, iv, 5, 11, 12, 14, 16, 17, 19, 40, 51, 55, 65, 69, 81, 82
conservation, 20, 61, 64, 69, 70, 76, 88
corpuscular aspect of light, 3
Cosmic Substratum, 3, 14, 15, 19, 22, 29, 33, 34, 40, 50, 65, 79, 80, 82, 85
discrepancy, 17, 18, 39, 78, 80-83
dogmas, iv
double star, 4
Earth, iv, 2, 10, 14, 21, 30, 49, 58-60, 75, 76, 78-82, 88-90
Einstein, iv, 1-4, 9, 12, 16, 20, 21, 33-36, 39, 48, 58, 59, 61, 65, 77, 84-89
electromagnetic waves, 2, 35
electromagnetism, 2, 60, 61, 90
energy, 2, 7, 14, 30, 31, 61-64, 69-76
energy conservation law, 20, 70
equivalence of all inertial frames, 3, 33, 36, 37
Fermat's principle, 91
four dimensional space, 12
fundamental frame, iv, 65, 75
fundamental theory, 31, 63-66, 72, 74-76, 80
Galilean law of addition of velocities, 1
Galilean relationships, v

Galilean transformations, iv, 20, 21, 29, 66
Galileo, 2, 20, 40, 59-61
group, v, 2, 3, 19, 33, 35-37, 60
group structure, v, 2, 3, 33, 35-37, 60
hidden variables, 69
inertial, iv, 1-3, 12, 15, 19-22, 29-31, 33-37, 39, 40, 44, 52, 59, 61, 62, 64-66, 69, 72-74, 76, 79
inertial frame, 15, 30, 40, 66, 72, 74, 76
Inertial systems, 86
Inertial transformations, 39, 40, 44
Invariance of the speed of light, 3
isotropic, 3, 4, 10, 11, 16, 19, 33, 35, 40, 42, 55, 89
isotropy, 21, 39, 48, 55, 87
kinetic energy, 30, 31, 62, 64, 73, 76
lack of synchronism, 81
law of addition of velocities, 1
law of composition of velocities, 2, 19, 20, 42-44, 80
length contraction, iv, 1, 5, 19, 21, 40, 44, 45, 47, 50, 51, 57, 58, 60, 66, 67, 69, 82, 92
Lorentz, iv, v, 1-5, 12, 15-17, 19-21, 29, 34-37, 39, 41, 42, 47, 48, 58, 60, 70, 77, 79, 81, 85, 86
Lorentz assumptions, 21
Lorentz transformations, 2, 60
Lorentz-Fitzgerald contraction, 47, 48
Lorentz-Poincaré transformations, iv, v, 2, 12, 15, 17, 19, 21, 29, 70
mass, 7, 19, 20, 43, 47, 61, 63, 64, 69-76
mass-energy, 62
mass-energy conservation, 20, 63
mass-energy equivalence, 64, 70
medium, 2, 3, 35, 60
mesons, 8, 14
Michelson interferometer, 50
Michelson's experiment, 1, 2, 47, 48, 90
microwave background, 49
Minkowski's space-time, 66
mirrors, 3, 8, 52, 58

From Galileo to Lorentz... and Beyond
Joseph Lévy (Montreal: Apeiron 2003)

momentum, 59, 70, 71
muon flux anisotropy, 49
one way speed of light, 4, 10, 48, 55, 58, 65, 77, 79, 87
photon, 2, 7
Poincaré, v, 1-4, 12, 13, 16, 19-21, 29, 33-36, 39, 42, 48, 59-61, 65, 66, 77, 84, 85
principle of inertia, 63, 64
quantity of motion, 69
real time, 11, 14, 17, 32, 42, 80, 82
reciprocity, 7, 61, 65, 75
reference frame, 3, 7, 9, 11, 15, 19, 22, 33, 34, 40-42, 44, 57, 66, 72-74
reference system, 33, 39, 59, 63, 88
Relativity of simultaneity, 66
relativity of time, 3, 65, 66
relativity postulate, 35, 60, 87
relativity principle, v, 1-3, 12-14, 19, 20, 29-31, 33, 35, 36, 59-63, 65, 66, 67, 74, 76, 85, 89, 90
relativity theory, i, ii, 1, 4, 5, 7, 20, 64, 65, 72, 75, 85, 87, 88, 89, 91, 92
rockets, 8
round trip, 77
slow clock transport, 4, 12, 16, 21, 48, 78, 79, 83, 87
space-ship, 65

space-time transformations, iv, v, 3, 5, 19-22, 29, 33, 35, 37, 66, 69
special relativity, 8, 9, 21, 35, 56, 64-66, 74, 78, 79, 85, 86, 88
Special relativity, 7, 87
speed of light, v, 1, 3, 4, 7, 8, 10, 11, 14-16, 19, 21, 33-35, 40, 48-52, 54-58, 62, 65, 74, 77-79, 81, 83-85, 87-90
synchronism discrepancy, 17, 18, 39, 78, 80-83
synchronization, iv, 5, 12, 13, 17, 19, 21, 39, 42, 65, 66, 69, 84
Synchronization, 77
systematic errors of measurement, 5, 12, 19, 29, 32, 33, 65, 66
theory of relativity, 7, 73, 86, 87, 89
toothed wheels experiment, 49, 58, 91
total available energy, 76
Twin paradox, 65
two way speed of light, 4, 21, 48, 50, 51, 52, 55-57, 65, 87-90
two way transit time of light, 4, 16, 50, 54-56, 77
velocity, v, 4, 5, 7, 8, 13, 14, 20, 29, 32, 39, 43, 48, 49, 64, 72, 77, 85, 87, 89

Made in the USA
Lexington, KY
07 October 2013